Student's Solutions Manual

James Lapp

Using & Understanding Mathematics: A Quantitative Reasoning Approach

Seventh Edition

Jeffrey Bennett
University of Colorado at Boulder

William Briggs
University of Colorado at Denver

Reproduced by Pearson from electronic files supplied by the author.

Publishing as Pearson, 501 Boylston Street, Boston, MA 02116.

1 18

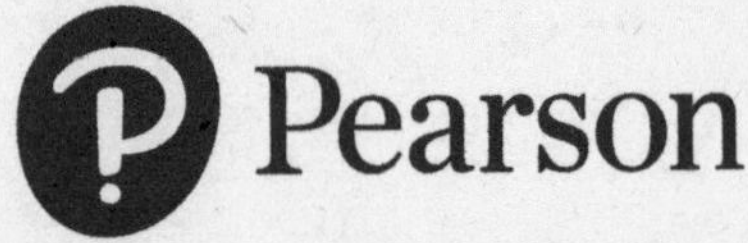

ISBN-13: 978-0-13-470524-8
ISBN-10: 0-13-470524-6

Table of Contents

UNIT 1A: LIVING IN THE MEDIA AGE

QUICK QUIZ

1. a. By the definition used in this book, an argument always contains at least one premise and a conclusion.
2. c. By definition, a fallacy is a deceptive argument.
3. b. An argument must contain a conclusion.
4. a. Circular reasoning is an argument where the premise and the conclusion say essentially the same thing.
5. b. Using the fact that a statement is unproved to imply that it is false is appeal to ignorance.
6. b. "I don't support the President's tax plan" is the conclusion because the premise "I don't trust his motives" supports that conclusion.
7. b. This is a personal attack because the premise (I don't trust his motives) attacks the character of the President, and says nothing about the substance of his tax plan.
8. c. This is limited choice because the argument does not allow for the possibility that you are a fan of, say, boxing.
9. b. Just because *A* preceded *B* does not necessarily imply that *A* caused *B*.
10. a. By definition, a straw man is an argument that distorts (or misrepresents) the real issue.

REVIEW QUESTIONS

1. Logic is the study of the methods and principles of reasoning. Arguing logically may or may not change a person's position, but it can give either person insight into the other's thinking.

3. A fallacy is an argument in which the conclusion is not well supported by the premises. The examples of fallacies given in the text are appeal to popularity, false cause, appeal to ignorance, hasty generalization, limited choice, appeal to emotion, personal attack, circular reasoning, diversion (red herring), and straw man. Examples will vary.

DOES IT MAKE SENSE?

5. Does not make sense. Raising one's voice has nothing to do with logical arguments.

7. Makes sense. A logical person would not put much faith in an argument that uses premises he believes to be false to support a conclusion.

9. Does not make sense. One can disagree with the conclusion of a well-stated argument regardless of whether it is fallacious.

BASIC SKILLS AND CONCEPTS

11. a. *Premise*: Apple's iPhone outsells all other smart phones. *Conclusion*: It must be the best smart phone on the market.

    b. The fact that many people buy the iPhone does not necessarily mean it is the best smart phone.

13. a. *Premise*: Decades of searching have not revealed life on other planets. *Conclusion*: Life in the universe must be confined to Earth.

    b. Failure to find life does not imply that life does not exist.

15. a. *Premise*: He refused to testify. *Conclusion*: He must be guilty.

    b. There are many reasons that someone might have for refusing to testify (being guilty is only one of them), and thus this is the fallacy of limited choice.

17. a. *Premise*: Senator Smith is supported by companies that sell genetically modified crop seeds. *Conclusion*: Senator Smith's bill is a sham.

    b. A claim about Senator Smith's personal behavior is used to criticize his bill.

19. a. *Premise*: Good grades are needed to get into college, and a college diploma is necessary for a good career. *Conclusion*: Attendance should count in high school grades.

    b. The premise (which is often true) directs attention away from the conclusion.

21. false; Explanations will vary.

23. false; Explanations will vary.

FURTHER APPLICATIONS

25. *Premise*: Eating oysters for dinner, followed by a nightmare. *Conclusion*: Oysters cause nightmares. This argument suffers from the **false cause** fallacy. We cannot conclude that the former caused the latter simply because they happened together.

27. *Premise*: All the nurses in Belvedere Hospital are women. *Conclusion*: Women are better qualified for medical jobs. The conclusion has been reached with a **hasty generalization**, because a small number of female nurses were used as evidence to support a claim about all men and women.

29. *Premise*: My uncle never drank alcohol and lived to be 93. *Conclusion*: Avoiding alcohol leads to greater longevity. **False cause** is at play here, as the abstinence from alcohol may have nothing to do with the uncle's longevity, even though both are occurring at the same time.

31. *Premise*: Five hundred million copies of *Don Quixote* have been sold. *Conclusion*: *Don Quixote* is popular. This is an **appeal to popularity**.

33. *Premise*: After I last gave to a charity, an audit showed that most of the money was used to pay its administrators in the front office. *Conclusion*: Charities cannot be trusted. The conclusion has been reached with a **hasty generalization**, because a small number charities are not passing donations on to the intended recipients does not mean that all charities do not pass on donations to the intended recipients.

35. *Premise*: The senator is a member of the National Rifle Association. *Conclusion*: I'm sure she opposes a ban on large capacity magazines. This is a **personal attack** on members of the National Rifle Association. The argument also distorts the position of the National Rifle Association (not all members would oppose a ban on large capacity magazines); this is also a **straw man**.

37. *Premise*: Some Democrats support doubling the federal minimum wage. *Conclusion*: Democrats think that everyone should have the same income. The argument distorts the position of the Democrats; this is a **straw man**.

39. *Premise*: My boy loves dolls, and my girl loves trucks. *Conclusion*: There's no truth to the claim that boys prefer mechanical toys while girls prefer maternal toys. Using one child of each gender to come up with a conclusion about all children is **hasty generalization**. It can also be seen as an **appeal to ignorance**: the lack of examples of boys enjoying mechanical toys (and girls maternal toys) does not mean that they don't enjoy these toys.

41. The example shows the fallacy of division because the fact that Jake is an American does not mean that he acts the same as all other Americans.

43. The example shows the fallacy of slippery slope because it assumes that the fact that troops have been sent to three countries means it's inevitable that they'll be sent to more.

UNIT 1B: PROPOSITIONS AND TRUTH VALUES

QUICK QUIZ

1. **c.** This is a proposition because it is a complete sentence making a claim, which could be true or false.

2. **a.** The truth value of a proposition's negation (*not p*) can always be determined by the truth value of the proposition.

3. **c.** Conditional statements are, by definition, in the form of *if p, then q*.

4. **c.** The table will require eight rows because there are two possible truth values for each of the propositions x, y, and z.

5. **c.** Because it is not stated otherwise, we are dealing with the inclusive *or* (and thus either p is true, or q is true, or both are true).
6. **a.** The conjunction p *and* q is true only when both are true, and since p is false, p *and* q must also be false.
7. **b.** This is the correct rephrasing of the original conjunction.
8. **c.** This is the *contrapositive* of the original conjunction.
9. **b.** Statements are logically equivalent only when they have the same truth values.
10. **a.** Rewriting the statement in *if* p, *then* q form gives, "if you want to win, then you've got to play."

REVIEW QUESTIONS

1. A proposition makes a claim (either an assertion or a denial) that may be either true or false. It must have the structure of a complete sentence. Examples will vary.
3. Given two propositions p and q, the statement p *and* q is called their conjunction, the statement p *or* q is called their disjunction, and a statement of the form *if* p, *then* q is called a conditional proposition (or implication). Examples will vary.
5. Truth tables for p *and* q, p *or* q, and *if* p, *then* q.

p	q	p *and* q
T	T	T
T	F	F
F	T	F
F	F	F

p	q	p *or* q
T	T	T
T	F	T
F	T	T
F	F	F

p	q	*if* p, *then* q
T	T	T
T	F	F
F	T	T
F	F	T

DOES IT MAKE SENSE?

7. Does not make sense. Propositions are never questions.
9. Makes sense. If restated in *if* p, *then* q form, this statement would read, "If we catch him, then he will be dead or alive." Clearly this is true, as it covers all the possibilities. (One could argue semantics, and say that a dead person is not caught, but rather discovered. Splitting hairs like this might lead one to claim the statement does not make sense).
11. Does not make sense. Not all statements fall under the purview of logical analysis.

BASIC SKILLS AND CONCEPTS

13. Since it's a complete sentence that makes a claim (whether true or false is immaterial), it's a proposition.
15. No claim is made with this statement, so it's not a proposition.
17. Questions are never propositions.
19. The negation is *Asia is not in the northern hemisphere.* The original proposition is true; the negation is false.
21. The negation is *The Beatles were a German band.* The original proposition is true; the negation is false.
23. Sarah did go to dinner.
25. Taxes will not be lowered.
27. Sue wants new trees planted in the park.

29. This is the truth table for the conjunction *q and r*.

q	*r*	*q and r*
T	T	T
T	F	F
F	T	F
F	F	F

31. "Dogs are animals" is true and "oak trees are plants" is also true. Since both propositions are true, the conjunction is true.

33. "Venus is a planet" is true and "The Sun is a star" is also true. Since both propositions are true, the conjunction is true.

35. "All birds can fly" is false and "some fish live in trees" is also false. The conjunction is false because both propositions in a conjunction must be true for the entire statement to be true.

37. This is the truth table for *q and r and s*.

q	*r*	*s*	*q and r and s*
T	T	T	T
T	T	F	F
T	F	T	F
T	F	F	F
F	T	T	F
F	T	F	F
F	F	T	F
F	F	F	F

39. The exclusive *or* is used here as it is unlikely that the proposition means you might both walk or ride a bike to the park.

41. The exclusive *or* is used here in the sense of which book will be read first.

43. The inclusive *or* is used here as you probably would be thrilled to both scuba dive or surf on your next vacation.

45. This is the truth table for the disjunction *r or s*.

r	*s*	*r or s*
T	T	T
T	F	T
F	T	T
F	F	F

47. This is the truth table for *p and (not p)*.

p	*not p*	*p and (not p)*
T	F	F
F	T	F

49. This is the truth table for *p or q or r*.

p	*q*	*r*	*p or q or r*
T	T	T	T
T	T	F	T
T	F	T	T
T	F	F	T
F	T	T	T
F	T	F	T
F	F	T	T
F	F	F	F

51. "Elephants are animals" is true and "elephants are plants" is false. The disjunction is true because a disjunction is true when at least one of its propositions is true.

53. Both "$3 \times 5 = 15$" and "$3 + 5 = 8$" are true. The disjunction is true because a disjunction is true when at least one of its propositions is true.

55. Both "Cars swim" and "dolphins fly" are false. The disjunction is false because a disjunction is false when all of its propositions are false.

57. This is the truth table for *if p, then r*.

p	r	*if p, then r*
T	T	T
T	F	F
F	T	T
F	F	T

59. *Hypothesis*: Trout can swim. *Conclusion*: Trout are fish. Both propositions are true, and the conditional proposition (implication) is true.

61. *Hypothesis*: Paris is in France. *Conclusion*: New York is in China. The hypothesis is true, the conclusion is false, and the conditional proposition is false.

63. *Hypothesis*: Trees can walk. *Conclusion*: Birds wear wigs. The hypothesis is false, the conclusion is false, and the conditional proposition is true.

65. *Hypothesis*: Dogs can swim. *Conclusion*: Dogs are fish. The hypothesis is true, the conclusion is false, and the conditional proposition is false.

67. If it snows (p), then I get cold (q).

69. If you are breathing (p), then you are alive (q).

71. If you are pregnant (p), then you are a woman (q).

73. *Converse*: If Tara owns a car, then she owns a Cadillac. Inverse: If Tara does not own a Cadillac, then she does not own a car. *Contrapositive*: If Tara does not own a car, then she does not own a Cadillac. The original proposition and the contrapositive are equivalent. The converse and inverse are equivalent.

75. *Converse*: If Helen is a U.S. citizen, then she is the U.S. President. Inverse: If Helen is not the U.S. President, then she is not a U.S. citizen. *Contrapositive*: If Helen is not a U.S. citizen, then she is not the U.S. President. The original proposition and the contrapositive are equivalent. The converse and inverse are equivalent.

77. *Converse*: If there is gas in the tank, then the engine is running. Inverse: If the engine is not running, then there is no gas in the tank. *Contrapositive*: If there is no gas in the tank, then the engine is not running. The original proposition and the contrapositive are equivalent. The converse and inverse are equivalent.

FURTHER APPLICATIONS

79. If you don't have passion, then you don't have energy. If you don't have energy, then you have nothing.

81. If you are excellent at flipping fries at McDonald's, then everyone will want to be in your line.

83. "If Sue lives in Cleveland, then she lives in Ohio," where it is assumed that Sue lives in Cincinnati. (Answers will vary.) Because Sue lives in Cincinnati, the hypothesis is false, while the conclusion is true, and this means the implication is true. The converse, "If Sue lives in Ohio, then she lives in Cleveland," is false, because the hypothesis is true, but the conclusion is false.

85. "If Ramon lives in Albuquerque, then he lives in New Mexico" where it is assumed that Ramon lives in Albuquerque. (Answers will vary.) The implication is true, because the hypothesis is true and the conclusion is true. The contrapositive, "If Ramon does not live in New Mexico, then he does not live in Albuquerque", is logically equivalent to the original conditional, so it is also true.

87. "If it is a fruit, then it is an apple." (Answers will vary.) The implication is false because, when the hypothesis is true, the conclusion may be false (it could be an orange). In the converse, "If it is an apple, then it is a fruit.", when the hypothesis is true, the conclusion is true, and this means the implication is true.

89. (a) Believing is sufficient for achieving.
(b) Achieving is necessary for believing.

91. (a) Having six children is sufficient for being committed.
(b) Being committed is necessary for having six children.

93. Following is a truth table for both *not* (*p and q*) and (*not p*) *or* (*not q*).

p	*q*	*p and q*	*not* (*p and q*)	(*not p*) *or* (*not q*)
T	T	T	F	F
T	F	F	T	T
F	T	F	T	T
F	F	F	T	T

Since both statements have the same truth values (compare the last two columns of the table), they are logically equivalent.

95. Following is a truth table for both *not* (*p and q*) and (*not p*) *and* (*not q*).

p	*q*	*p and q*	*not* (*p and q*)	(*not p*) *and* (*not q*)
T	T	T	F	F
T	F	F	T	F
F	T	F	T	F
F	F	F	T	T

Note that the last two columns in the truth table don't agree, and thus the statements are not logically equivalent.

97. Following is a truth table for (*p and q*) *or r* and (*p or r*) *and* (*p or q*).

p	*q*	*r*	*p and q*	(*p and q*) *or r*	*p or r*	*p or q*	(*p or r*) *and* (*p or q*)
T	T	T	T	T	T	T	T
T	T	F	T	T	T	T	T
T	F	T	F	T	T	T	T
T	F	F	F	F	T	T	T
F	T	T	F	T	T	T	T
F	T	F	F	F	F	T	F
F	F	T	F	T	T	F	F
F	F	F	F	F	F	F	F

Since the fifth and eighth column of the table don't agree, these two statements are not logically equivalent.

99. Given the implication *if p, then q* the contrapositive is *if* (*not q*) *then* (*not p*). The converse is *if q, then p* and the inverse of the converse is *if* (*not q*) *then* (*not p*), which is the contrapositive. Similarly, the contrapositive is also the converse of the inverse.

UNIT 1C: SETS AND VENN DIAGRAMS

QUICK QUIZ

1. b. The ellipsis is a convenient way to represent all the other states in the U.S. without having to write them all down.
2. c. $3\frac{1}{2}$ is a rational number (a ratio of two integers), but it is not an integer.
3. a. When the circle labeled C is contained within the circle labeled D, it indicates that C is a subset of D.

4. **b.** Since the set of cats is disjoint from the set of dogs, the two circles should be drawn as non-overlapping circles.

5. **a.** Because all apples are fruit, the set *A* should be drawn within the set B (the set of apples is a subset of the set of fruits).

6. **c.** Some cross country runners may also be swimmers, so their sets should be overlapping.

7. **a.** The X is placed in the region where *business executives* and *working mothers* overlap to indicate that there is at least one member in that region.

8. **c.** The region X is within both *males* and *athletes*, but not within *Republicans*.

9. **a.** The central region is common to all three sets, and so represents those who are male, Republican, and an athlete.

10. **c.** The sum of the entries in the column labeled Low Birth Weight is 32.

REVIEW QUESTIONS

1. A set is a collection of objects. Sets are often described by listing their members within a pair of braces, { }.

3. All *S* are *P*.

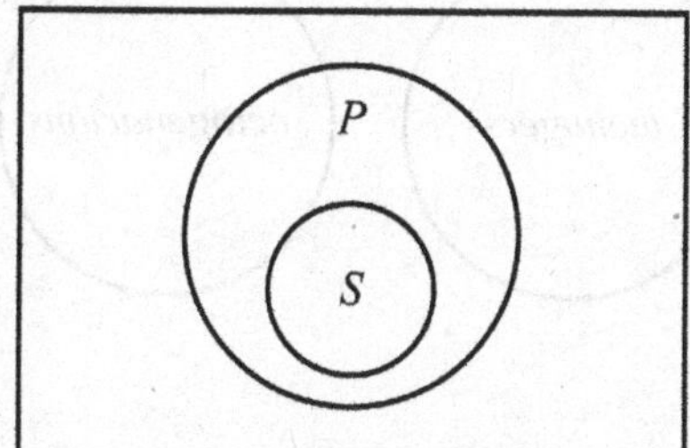

Some *S* are *P*.

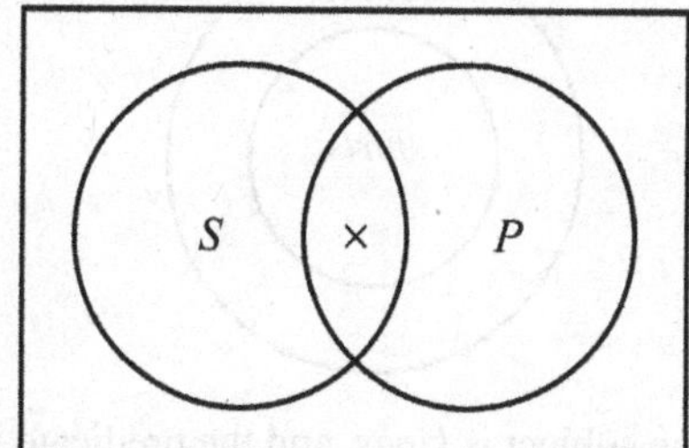

No *S* are *P*.

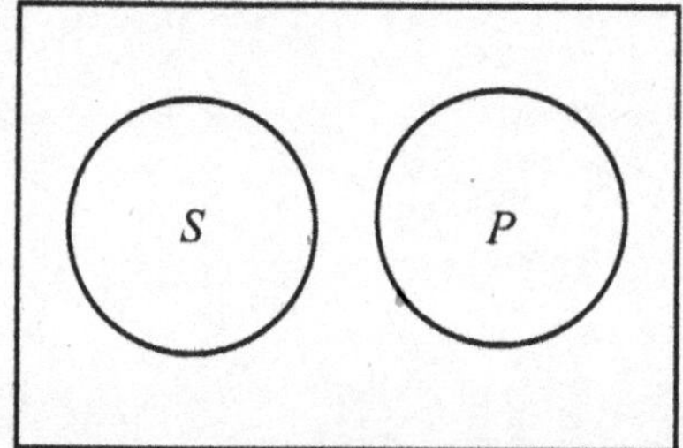

Some *S* are not *P*.

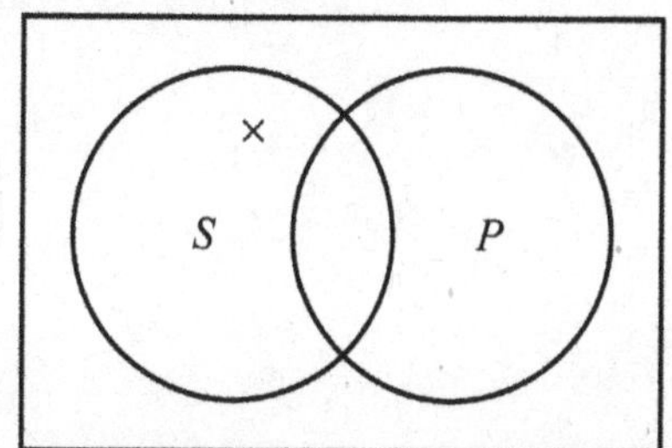

5. Draw three overlapping circles (As shown in Figure 1.21.). Each of the eight regions represent a distinct combination of the three properties indicated by the three circles.

DOES IT MAKE SENSE?

7. Does not make sense. There are people that live in houses in Chicago, and would not be an element of the set of those who rent apartments.

9. Does not make sense. The number of students in a class is a whole number, and whole numbers are not in the set of irrational numbers.

11. Does not make sense. A Venn diagram shows only the relationship between members of sets, but does not have much to say about the truth value of a categorical proposition.

BASIC SKILLS AND CONCEPTS

13. 888 is a natural number.

15. 3/4 is a rational number.

17. 3.414 is a rational number.

19. π is a real number.

21. −45.12 is a rational number.

23. $\pi/4$ is a real number.

25. $-123/79$ is a rational number.

27. $\pi/129$ is a real number.

29. {1, 2, 3, …, 30, 31}

31. {Alabama, Arkansas, Louisiana, Tennessee}

33. {16, 25, 35}

35. {6, 12, 18, 24, 30}

37. Because some women are teachers, the circles should overlap.

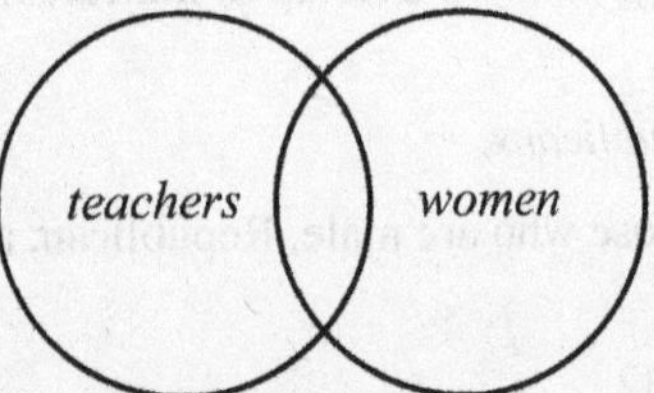

39. Shirts are clothing, and thus the set of shirts is a subset of the set of clothing. This means one circle should be contained within the other.

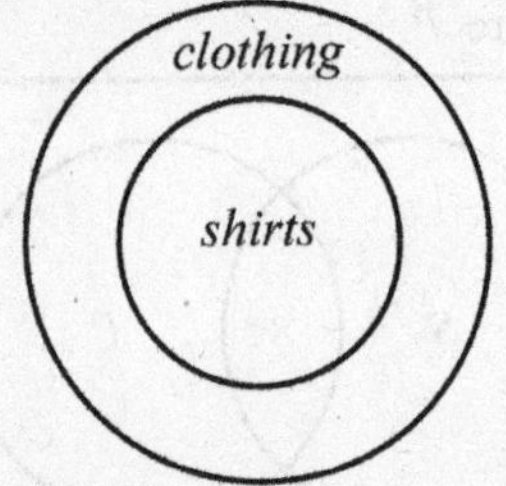

41. Some poets are also plumbers, so the circles should overlap.

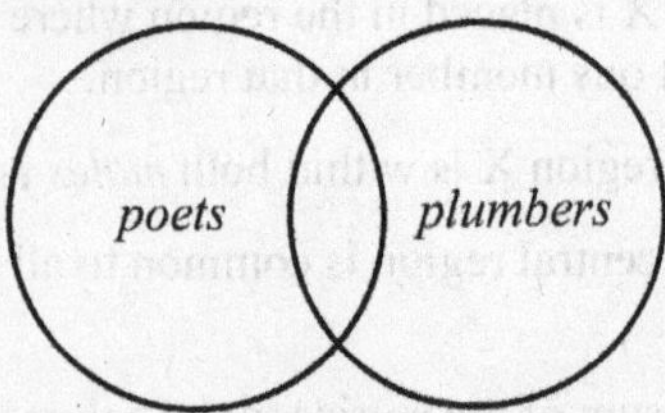

43. No teenager is an octogenarian, so these sets are disjoint, and the circles should not overlap.

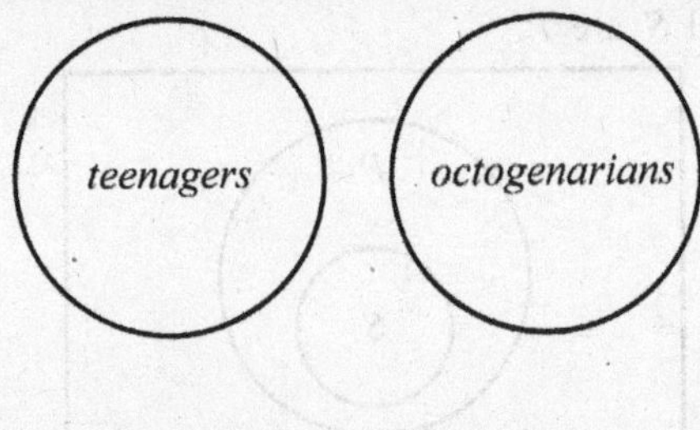

45. b. The subject is *kings*, and the predicate is *men*.

c.

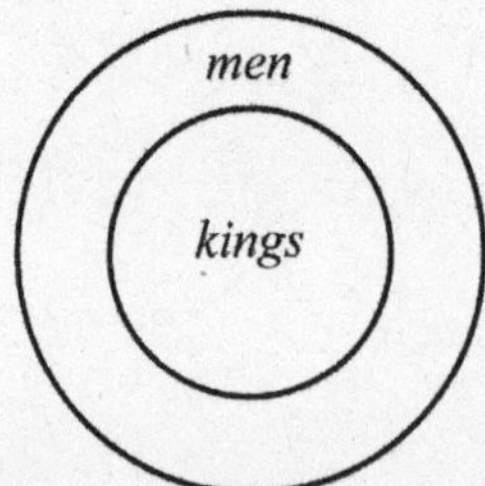

d. No, the diagram does not show evidence that there is a king that is not a man.

47. b. The subject is *surgeons*, and the predicate is *fisherman*.

c.

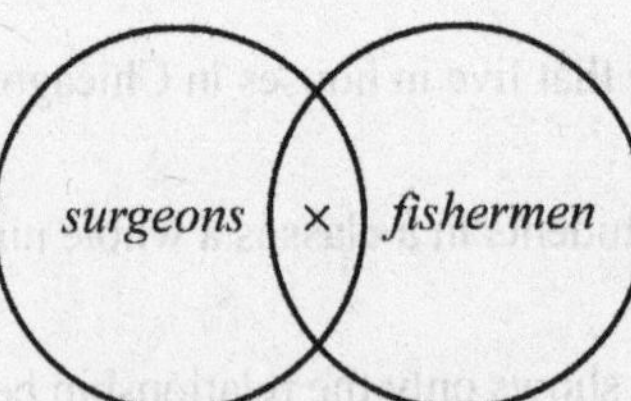

d. No, the diagram gives no evidence that there are fishermen that are not surgeons.

49. a. No monks swear.

b. The subject is *monks*, and the predicate is *swearers*.

c.

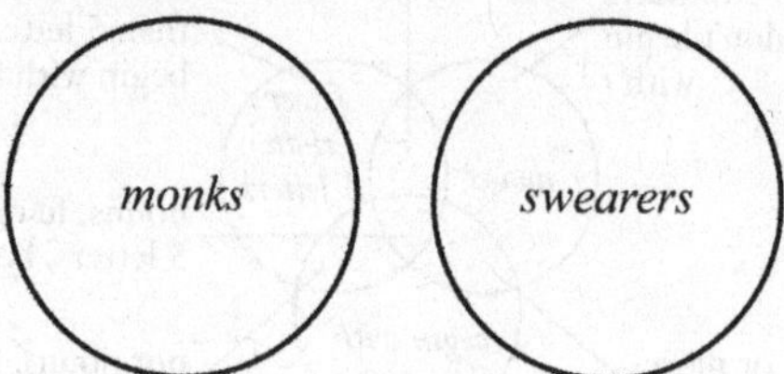

d. No, since the sets are disjoint, the would have no common members.

51. b. The subject is *sharpshooters*, and the predicate is *men*.

c.

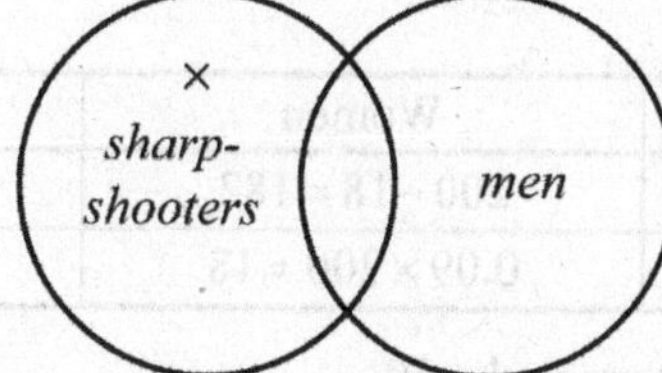

d. No, the diagram gives no evidence that there is a sharpshooter that is a man.

53.

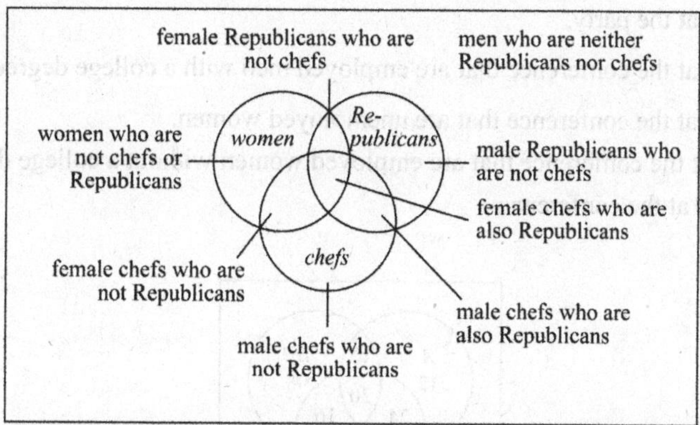

55.

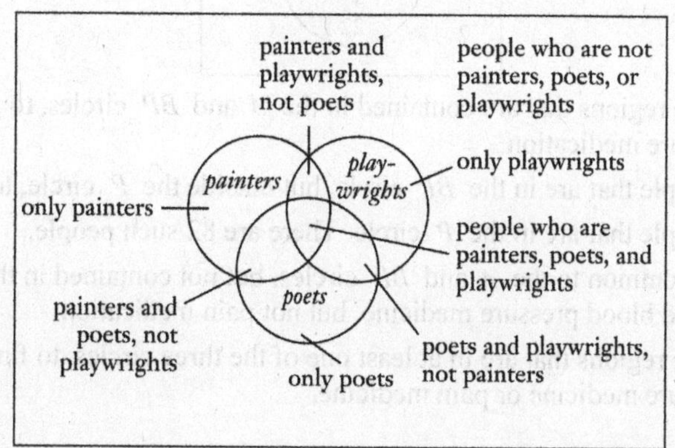

57.

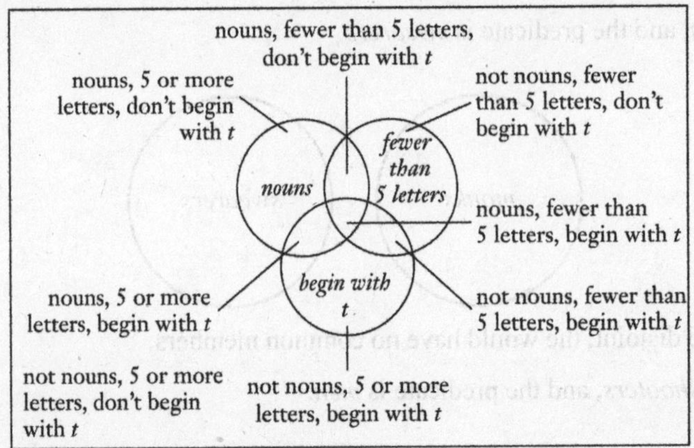

59.

	Women	Men
Right-handed	$200-18=182$	$150-18=132$
Left-Handed	$0.09\times 200=18$	$0.12\times 150=18$

61. a. There are 31 people at the party that are under 30.
b. There are 28 women at the party that are not under 30.
c. There are 37 men at the party.
d. There are 81 people at the party.

63. a. There are 16 people at the conference that are employed men with a college degree.
b. There are 24 people at the conference that are unemployed women.
c. There are 8 people at the conference that are employed women without a college degree.
d. There are 43 women at the conference.

65. a.

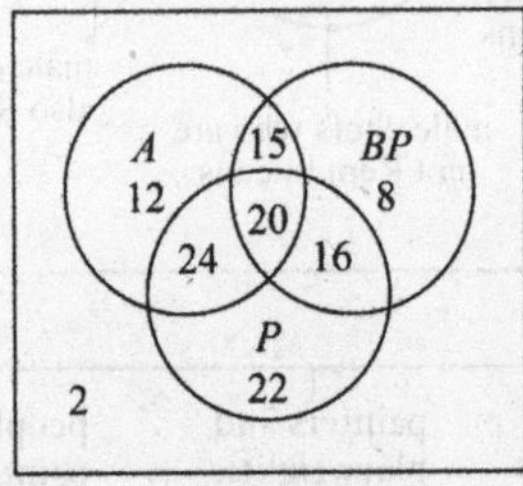

b. Add the numbers in the regions that are contained in the A and BP circles, to find that 95 people took antibiotics or blood pressure medication.
c. Add the number of people that are in the BP circle, but outside the P circle, to arrive at 23 people.
d. Add the number of people that are in the P circle. There are 82 such people.
e. Use the region that is common to the A and BP circles, but not contained in the P circle, to find that 15 people took antibiotics and blood pressure medicine, but not pain medication.
f. Add the numbers in the regions that are in at least one of the three circles, to find that 117 people took antibiotics or blood pressure medicine or pain medicine.

FURTHER APPLICATIONS

67. a.

	Favorable Review	Non-favorable Review	Total
Documentaries	8	$12-8=4$	12
Feature Films	$24-6=18$	6	24
Total	$8+18=26$	$4+6=10$	36

b.

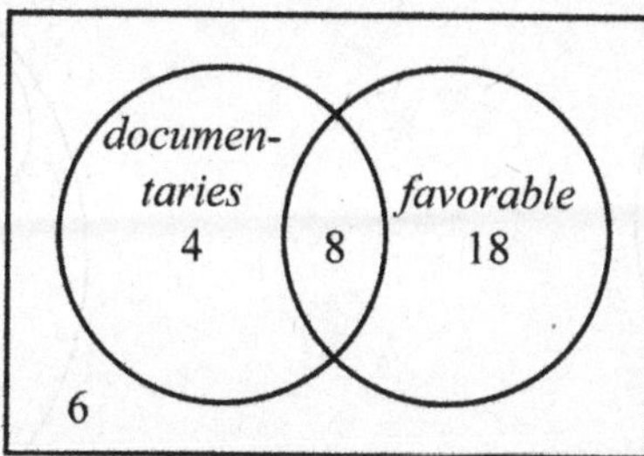

c. 4 documentaries received unfavorable reviews.

d. 18 feature films received favorable reviews.

69. a.

	Blues	Country	Total
Nashville	$51-35=16$	35	$100-49=51$
San Francisco	30	19	$30+19=49$
Total	$16+30=46$	$35+19=54$	100

b.

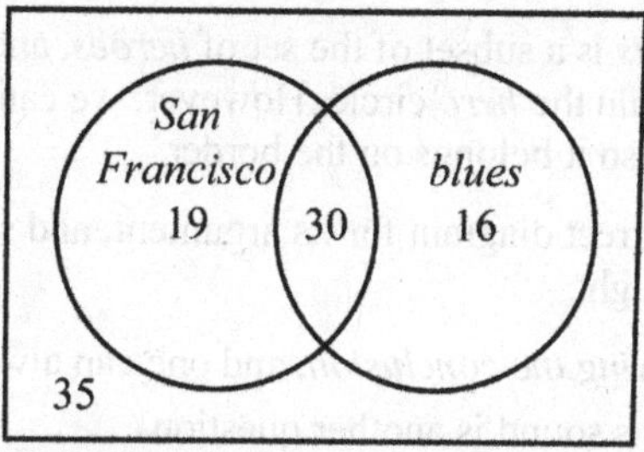

c. 16 Nashville respondents preferred blues.

d. 46 respondents preferred blues.

71. Answers will vary.

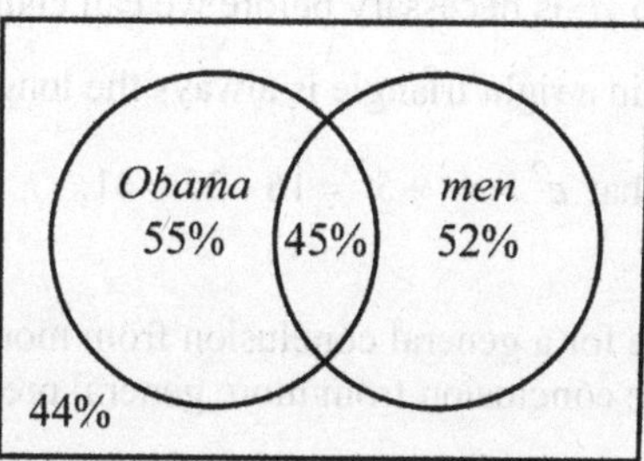

73. a. The two-way table should look like this:

	Subject did not lie.	Subject lied.	Total
Polygraph: lie	$57-42=15$	42	57
Polygraph: no lie	32	$51-42=9$	$32+9=41$
Total	$15+32=47$	51	$57+41=98$

73 (continued)

b. $\frac{42+32}{98}=0.755=75.5\%$

c. $\frac{15+9}{98}=0.245=24.5\%$

d. Answers will vary.

75.

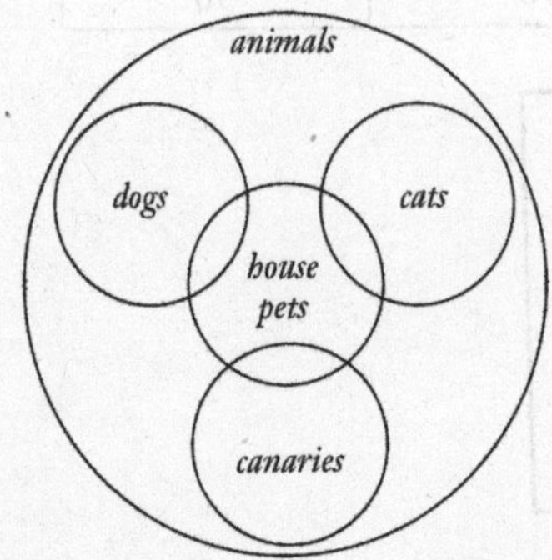

77.

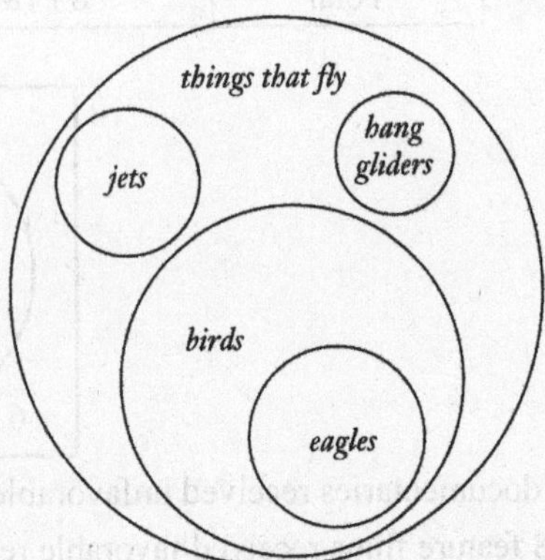

UNIT 1D: ANALYZING ARGUMENTS

QUICK QUIZ

1. **b.** The only way to prove a statement true beyond all doubt is with a valid and sound deductive argument.
2. **c.** A deductive argument that is valid has a logical structure that implies its conclusion from its premises.
3. **c.** If a deductive argument is not valid, it cannot be sound.
4. **a.** Premise 1 claims the set of *knights* is a subset of the set of *heroes*, and Premise 2 claims Paul is a hero, which means the X must reside within the *hero* circle. However, we cannot be sure whether the X should fall within or outside the *knights* circle, so it belongs on the border.
5. **c.** Diagram *a* in question 4 is the correct diagram for its argument, and since X lies on the border of the *knights* circle, Paul may or may not be a knight.
6. **b.** The argument is of the form *denying the conclusion*, and one can always conclude p is not true in such arguments. (Whether the argument is sound is another question).
7. **c.** This argument is of the form *affirming the conclusion*, and it is always invalid, which means we can conclude nothing about p.
8. **c.** A chain of conditionals from a to d is necessary before we can claim the argument is valid.
9. **b.** The side opposite the right angle in a right triangle is always the longest, and it's called the hypotenuse.
10. **b.** The Pythagorean theorem states that $c^2=4^2+5^2=16+25=41$.

REVIEW QUESTIONS

1. An inductive argument makes a case for a general conclusion from more specific premises, while a deductive argument makes a case for a specific conclusion from more general premises.

3. An argument is valid if its conclusion follows necessarily from its premises, regardless of the truth of those premises or conclusions. An argument is sound if it is valid and its premises are all true, otherwise it is unsound. By definition, sound deductive arguments must be valid.

5. Examples will vary. The four basic conditional arguments are: Affirming the Hypothesis, Affirming the Conclusion, Denying the Hypothesis, and Denying the Conclusion.

7 Inductive logic cannot be used to prove a mathematical theorem. No matter how many true cases are provided, a false case my be just around the corner.

DOES IT MAKE SENSE?

9. Does not make sense. One cannot prove a conclusion beyond all doubt with an inductive argument.

11. Makes sense. As long as the logic of a deductive argument is valid, if one accepts the truth (or soundness) of the premises, the conclusion necessarily follows.

13. Does not make sense. This argument is of the form *affirming the conclusion*, and it is always invalid.

BASIC SKILLS AND CONCEPTS

15. This is an inductive argument because it makes the case for a general conclusion based on many specific observations.

17. This is an inductive argument because it makes the case for a general conclusion based on many specific observations.

19. This is a deductive argument because a specific conclusion is deduced from more general premises.

21. This is a deductive argument because a specific conclusion is deduced from more general premises.

23. The premises are true, though the argument is weak (it speaks to only six of many painters). The conclusion is false.

25. The premises are true, the argument *seems* moderately strong, and the conclusion is false.

27. The premises are true and the argument is moderately strong. The conclusion is correct.

29. a. The argument is valid.

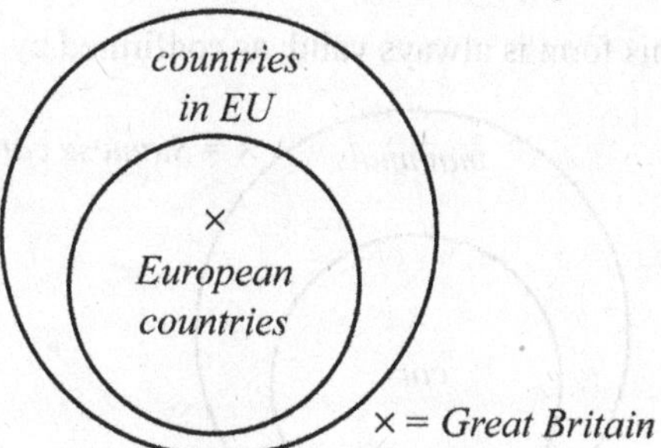

b. The argument is not sound since the first premise is false.

31. a. The argument is not valid.

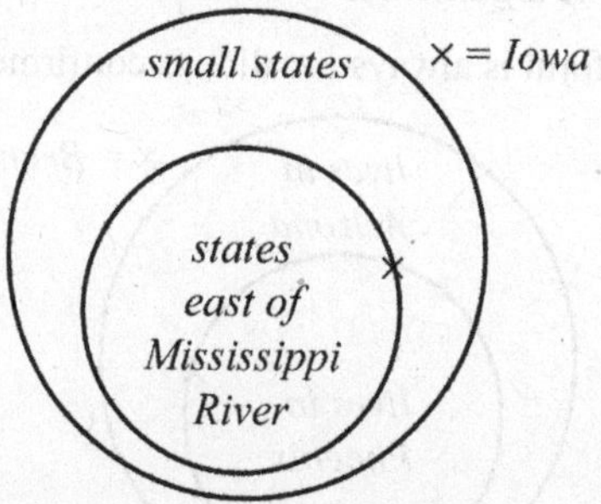

b. The diagram shows the argument is invalid, even though the premises are questionable. Because it is invalid, the argument cannot be sound.

33. a. The argument is valid.

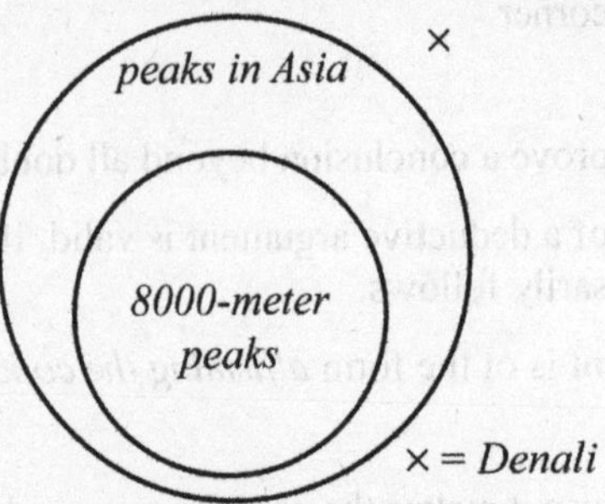

b. The premises are true, and the argument is sound.

35. a. The argument is not valid.

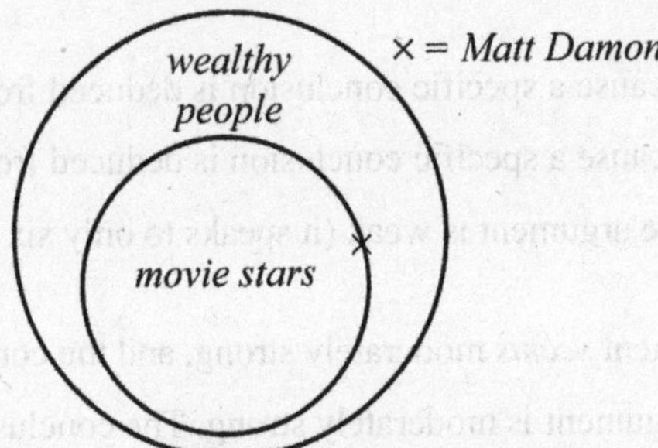

b. The diagram shows the argument is invalid, so the argument cannot be sound.

37. a. Affirming the hypothesis – this form is always valid, as confirmed by the diagram.

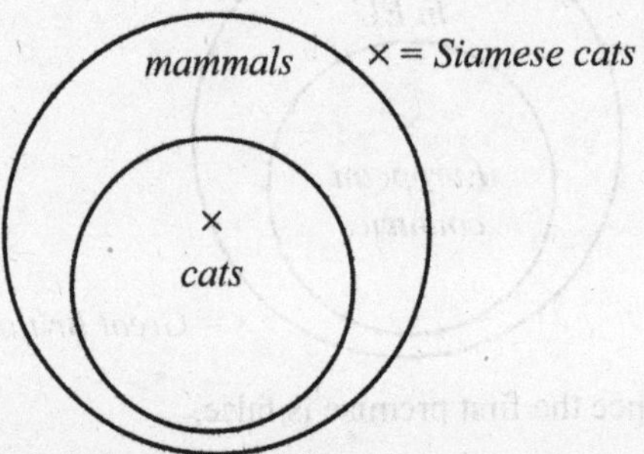

b. The premises are true, and thus the argument is sound.

39. a. Affirming the conclusion – this form is always invalid, as confirmed by the diagram.

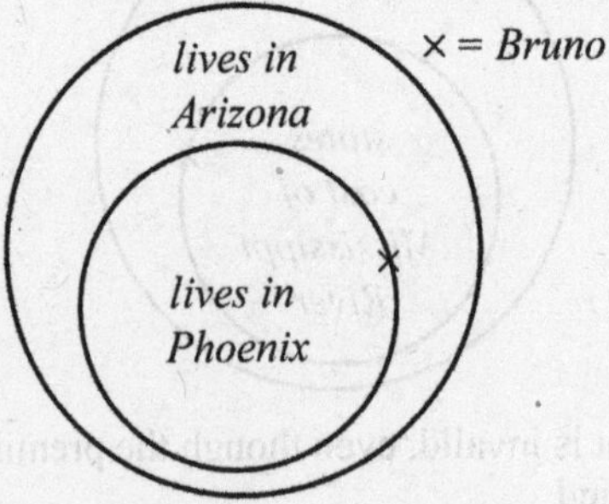

b. Since it is invalid, the argument cannot be sound.

41. a. Denying the hypothesis – this form is always invalid, as confirmed by the diagram.

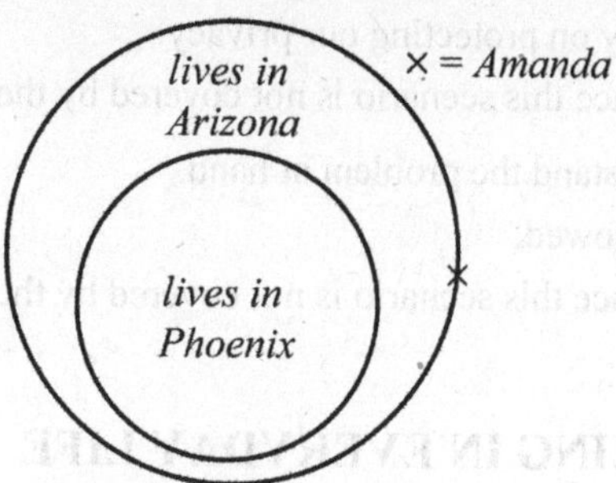

b. Since it is invalid, the argument cannot be sound.

43. a. Denying the conclusion – this form is always valid, as confirmed by the diagram.

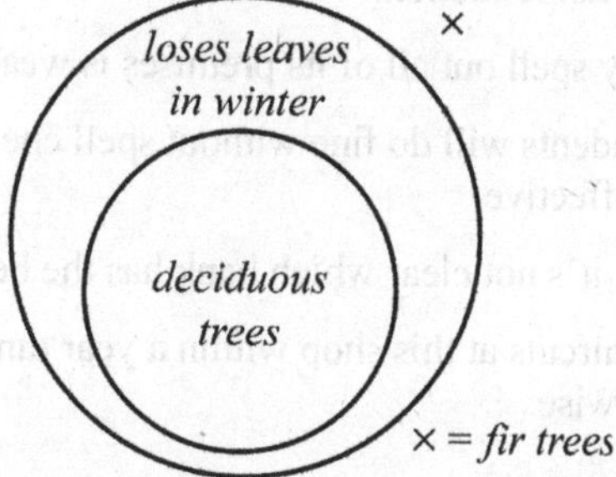

b. The premises are true, and thus the argument is sound.

45. $p =$ a natural number is divisible by 18, $q =$ a natural number is divisible by 9, $r =$ a natural number is divisible by 3; There is a clear chain of implications from the first premise to the conclusion, so the argument is valid.

47. There is a clear chain of implications from the first premise to the conclusion, so the argument is valid. To be valid, the conclusion should be: "If everyone is obeying the golden rule, then the world is more peaceful."

49. The statement is true.

51. The statement is not true. Counterexamples will vary. One possibility is: $\sqrt{9+16} = \sqrt{25} = 5$; $\sqrt{9} + \sqrt{16} = 3 + 4 = 7$; but $5 \neq 7$.

FURTHER APPLICATIONS

53. It is possible.
Answers will vary. An example:
Premise: All living mammals breathe.
Premise: All monkeys are mammals.
Conclusion: All living monkeys breathe.

55. It is possible.
Answers will vary. An example:
Premise: All mammals fly. (false)
Premise: All monkeys are mammals. (true)
Conclusion: All monkeys fly. (false)

57. It is possible.
Answers will vary. An example:
Premise: All mammals breathe. (true)
Premise: All mammals have hair. (true)
Conclusion: All hairy animals breathe. (true)

59. An example of affirming the hypothesis (valid):
Premise: If I am in Phoenix, then I am in Arizona.
Premise: I am in Phoenix.
Conclusion: I am in Arizona.

61. An example of denying the hypothesis (invalid):
Premise: If I am in Phoenix, then I am in Arizona.
Premise: I am not in Phoenix.
Conclusion: I am not in Arizona.

63. a. We will sacrifice both free enterprise and security.
   b. We did not insist too adamantly on protecting our privacy.
   c. No conclusion can be made since this scenario is not covered by the premises.

65. a. Apes act as if they don't understand the problem at hand.
   b. The usual procedures were followed.
   c. No conclusion can be made since this scenario is not covered by the premises.

67. Answers will vary.

UNIT 1E: CRITICAL THINKING IN EVERYDAY LIFE

QUICK QUIZ

1. **b.** This is the definition of critical thinking.
2. **b.** A vote for C implies a property tax reduction.
3. **c.** An argument that doesn't clearly spell out all of its premises is weak in logical structure.
4. **a.** The teacher is assuming that students will do fine without spell checkers, which implies that traditional methods of teaching spelling are effective.
5. **c.** With unknown application fees, it's not clear which bank has the better offer.
6. **b.** It's a good deal if you get six haircuts at this shop within a year (and that you remember to get your card punched), but it's a bad deal otherwise.
7. **c.** \$20/100 min = 20¢/min.
8. **a.** As long as you remember to get the 50% refund coming to you, you'll spend \$200.
9. **b.** You can't compute how much you'll spend with each policy without knowing the number and cost of collisions over the span of a year.
10. **c.** If it did not rain, and today is a Saturday, the Smiths would have a picnic. Since they did not, it must not be a Saturday.

REVIEW QUESTIONS

1. Critical thinking is the careful evaluation of evidence and arguments.

3. Answers will vary.

DOES IT MAKE SENSE?

5. Makes sense. The double negative means the insurance company accepted his claim.
7. Does not make sense. If Sue wants to save time, she should take the Blue Shuttle, and save ten minutes.
9. Makes sense. Both the duration and mileage of the first warranty is the better deal.

BASIC SKILLS AND CONCEPTS

11. A "yes" vote is a vote *against* gay rights, while a "no" vote is a vote *for* gay rights.

13. a. Yes, if it has a picture.
   b. No, it doesn't have a picture.
   c. Apply for a voter ID or sign a sworn statement and cast a provisional ballot.
   d. The initiative does not explain how to get a voter ID.

15. (1) Buying a house will continue to be a good investment. (2) You will spend less out-of-pocket on your home payments than you would on rent.

17. (1) The Governor will keep his promise on tax cuts. (2) you consider tax cuts to be more important than other issues.

19. We are looking for possible unstated motives that may be the unstated "real reason" for opposition to the spending proposal. Among the possibilities is that the speaker may have a fundamental ideological opposition to paying taxes.

21. Plan A costs $2200 if you go and $0 if you cancel. Plan B costs $1200 if you go and $300 if you cancel. Plan A costs $1000 more if you go. Plan B costs $300 more if you cancel. If the likelihood of cancellation is low then Plan B is better, but if the likelihood of cancellation is high then Plan A is better.

23. A legitimate sweepstakes would not ask you to pay a processing fee in order to claim your prize. Note also that the notice never says your vacation will be fully paid for. In addition, the notice is asking for your credit card number, which should raise a "red flag" that should cause you to delete the message as spam.

25. Pyramidologists will use real things, such as the way ancient structures often follow astronomical alignments, to support their belief that ancient people had deep knowledge, then simply assume that much more must be hidden, even though there is no evidence of it.

27. Popes have never taken a position in American politics. Of the two major candidates in 2016, one might question why the pope would support Trump's positions on a variety of issues.

FURTHER APPLICATIONS

29. He has 4 bagels left as he ate all but 4.

31. Neither person, roosters don't lay eggs.

33. You must meet 22 people, as the first twenty might all be Canadians.

35. You must meet three people, as the first two might be different nationalities.

37. Suzanne might go bowling 1, 2, 3, or 4 days per week.

39. No, it does not follow. All of the chocolate lovers may be men.

41. Under your current policy (and over the span of a nine-month pregnancy), you'll spend $115 per month, plus $4000 for prenatal care and delivery, for a total of $5035. Under the upgraded policy, you'll spend $275 per month, for a total of $2475. Thus, considering only these costs (we aren't told, for example, what happens if the mother requires an extensive hospital stay due to a C-section or other complications arising from delivery, or what happens if the baby is born prematurely and requires neo-natal care), the upgraded policy is best.

43. If you fly twenty-four times with Airline A, you will earn four free flights (because you will have flown 72,000 miles at that point). Your cost will be $20 \times \$350 = \$7000$. If you fly twenty-four times with Airline B, your cost will be $24 \times \$325 = \$7800$. Thus, Airline A is cheaper than Airline B.

45. a. Maria must file a return since her earned income is greater than $5950.

b. Van must file a return since his gross income of $3500 is greater than his earned income plus $300

c. Walt need not file a return since his income does not meet any of the criteria.

d. Helena must to file a return since her gross income of $6000 is greater than her earned income (up to $5650) plus $300 ($\$6000 > \$5650 + \$300$).

47. a. The minimum payment will be $45, the total of the $35 in unpaid late fees and $10, since it is greater than 5% of your new balance.

b. Yes, to avoid a finance charge, you must pay within 25 days of the statement closing date.

c. You will be assessed a finance charge since you still have an unpaid balance more than 25 days after the statement closing date

49. a. Accepting a campaign contribution from someone you have never met conforms with the law.

b. Accepting a contribution from a government campaign fund would conform with the law. Accepting money from a CEO who will benefit from a bill you are sponsoring would violate the law.

51–53. Answers will vary.

55. One interpretation of the poem is that the poet was 20 years old when he wrote the poem, and he expects to live to age 70.

UNIT 2A: UNDERSTAND, SOLVE, AND EXPLAIN

QUICK QUIZ

1. **b.** Speed is described by distance per unit of time, so dividing a distance by a time is the correct choice.
2. **a.** Think of the unit *miles per hour*; the unit of *mile* is divided by the unit of *hour*.
3. **b.** "Of" implies multiplication.
4. **a.** Take dollars per gallon and divide it by miles per gallon, and you get $\frac{\$}{\text{gal}} \div \frac{\text{mi}}{\text{gal}} = \frac{\$}{\text{gal}} \times \frac{\text{gal}}{\text{mi}} = \frac{\$}{\text{mi}}$.
5. **b.** The area of a square is its length multiplied by its width (these are, of course, equal for squares), and thus a square of side length 3 mi has area of $3 \text{ mi} \times 3 \text{ mi} = 9 \text{ mi}^2$.
6. **c.** When multiplying quantities that have units, the units are also multiplied, so $\text{ft}^2 \times \text{ft} = \text{ft}^3$.
7. **b.** $1 \text{ mi}^3 = (1760 \text{ yd})^3 = 1760^3 \text{ yd}^3$
8. **c.** $1 \text{ ft}^2 = 12 \text{ in} \times 12 \text{ in} = 144 \text{ in}^2$
9. **a.** Apples are most likely to be sold by units of weight (or more accurately, mass), and thus euros per kilogram is the best answer.
10. **b.** $1.058 per euro means 1 euro = $1.058, which is more than $1.

REVIEW QUESTIONS

1. See the Information Box on Page 72. Examples will vary.

3. *Per* means "for every," *of* implies multiplication, *square* implies the second power, and *cube* implies the third power. A hyphen represents the product of different units.

5. A square yard is three feet wide by three feet high, so it contains $3 \text{ ft} \times 3 \text{ ft} = 9 \text{ ft}^2$. A cubic yard is three feet wide by three feet high by three feet deep, so it contains $3 \text{ ft} \times 3 \text{ ft} \times 3 \text{ ft} = 27 \text{ ft}^3$. To find the conversion factor for squares and cubes, we square or cube the equivalent measure of the unit not raised to an exponent. In the preceding example, to find square yards and cubic yards, the number of feet in a yard was squared and cubed, respectively.

DOES IT MAKE SENSE?

7. Does not make sense. 35 miles is a distance, not a speed.

9. Makes sense. Using the familiar formula $distance = rate \times time$, one can see that dividing the distance by the rate (or speed) will result in time.

11. Does not make sense. $\frac{\$3}{1 \text{ ft}^2} \times \frac{9 \text{ ft}^2}{1 \text{ yd}^2} = \$27/\text{yd}^2$, which is more than the price at Y-mart.

BASIC SKILLS AND CONCEPTS

13. a. $\frac{3}{4} \times \frac{1}{2} = \frac{3 \cdot 1}{4 \cdot 2} = \frac{3}{8}$

b. $\frac{2}{3} \times \frac{3}{5} = \frac{2 \cdot 3}{3 \cdot 5} = \frac{2}{5}$

c. $\frac{1}{2} + \frac{3}{2} = \frac{1+3}{2} = \frac{4}{2} = 2$

d. $\frac{2}{3} + \frac{1}{6} = \frac{4}{6} + \frac{1}{6} = \frac{4+1}{6} = \frac{5}{6}$

e. $\frac{2}{3} \times \frac{1}{4} = \frac{2 \cdot 1}{3 \cdot 4} = \frac{2}{12} = \frac{1}{6}$

f. $\frac{1}{4} + \frac{3}{8} = \frac{2}{8} + \frac{3}{8} = \frac{2+3}{8} = \frac{5}{8}$

g. $\frac{5}{8} - \frac{1}{4} = \frac{5}{8} - \frac{2}{8} = \frac{5-2}{8} = \frac{3}{8}$

h. $\frac{3}{2} \times \frac{2}{3} = \frac{3 \cdot 2}{3 \cdot 2} = 1$

15. Answers may vary depending on whether fractions are reduced.

a. $3.5 = \frac{35}{10} = \frac{7}{2}$

b. $0.3 = \frac{3}{10}$

c. $0.05 = \frac{5}{100} = \frac{1}{20}$

d. $4.1 = \frac{41}{10}$

e. $2.15 = \frac{215}{100} = \frac{43}{20}$

f. $0.35 = \frac{35}{100} = \frac{7}{20}$

g. $0.98 = \frac{98}{100} = \frac{49}{50}$

h. $4.01 = \frac{401}{100}$

17. a. $\frac{1}{4} = 0.25$

b. $\frac{3}{8} = 0.375$

c. $\frac{2}{3} \approx 0.667$

d. $\frac{3}{5} = 0.6$

e. $\frac{13}{2} = 6.5$

f. $\frac{23}{6} \approx 3.833$

g. $\frac{103}{50} = 2.06$

h. $\frac{42}{26} \approx 1.615$

19. $28\text{ ft}^3 \times \frac{64.2\text{ lb}}{1\text{ ft}^3} = 1797.6\text{ lb}$

21. $2.5\text{ lb} \times \frac{\$1.25}{1\text{ lb}} = \$3.125$

23. $\frac{\$420}{23\text{ hours}} = \$18.26\text{ \$/hr}$

25. a. The area of the storage pod's floor is $12\text{ ft} \times 20\text{ ft} = 240\text{ ft}^2$, and the volume of the storage pod is $12\text{ ft} \times 20\text{ ft} \times 8\text{ f} = 1920\text{ ft}^3$.

b. The surface area of the pool is $25\text{ yd} \times 20\text{ yd} = 500\text{ yd}^2$, and the volume of water it holds is $25\text{ yd} \times 20\text{ yd} \times 2\text{ yd} = 1000\text{ yd}^3$.

c. The area of the bed is $30\text{ ft} \times 6\text{ ft} = 180\text{ ft}^2$, and the volume of soil it holds is $30\text{ ft} \times 6\text{ ft} \times 1.2\text{ ft} = 216\text{ ft}^3$.

27. Speed has units of miles per hour, or mi/hr.

29. The flow rate has units of cubic feet per second, or cfs (or ft^3/s).

31. The cost of a car trip has units of dollars, or \$.

33. Convert 1.2 cubic yards to dollars by using the conversion 1 cubic yard = \$24: $1.2\text{ yd}^3 \times \frac{\$24}{\text{yd}^3} = \$28.80$.

35. Convert 2.5 ounces to dollars by using the conversion 1 oz = \$1200: $2.5\text{ oz} \times \frac{\$1200}{\text{oz}} = \$3000$.

37. First, note that 321 million people is 3210 groups of 100,000 people each. The mortality rate is

$$\frac{595,700\text{ deaths}}{321,000,000\text{ people}} \times \frac{321,000,000\text{ people}}{3210\text{ groups of }100,000} = 185.6\text{ deaths/100,000 people}..$$

39. The cost to drive 30 miles is $30\text{ mi} \times \frac{1\text{ gal}}{32\text{ mi}} \times \frac{\$2.55}{1\text{ gal}} = \$2.39$.

41. The solution is wrong. It's helpful to include units on your numerical values. This would have signaled your error, as illustrated: (*incorrect*) $0.11\text{ lb} \div \frac{\$7.70}{\text{lb}} = \frac{0.11\text{ lb}^2}{\$7.70} = 0.014\frac{\text{lb}^2}{\$}$. Notice that your solution, when the correct units are added, produces units of square pounds per dollar, which isn't very helpful, and doesn't answer the question. The correct solution is $0.11\text{ lb} \times \frac{\$7.70}{\text{lb}} = \$0.85$.

43. The solution is wrong. Units should be included with all quantities. When dividing by \$11, the unit of dollars goes with the 11 into the denominator (as shown below). Also, while it's reasonable to round an answer to the nearest tenth, it's more useful to round to the nearest hundredth in a problem like this, as you'll be comparing the price per pound of the large bag to the price per pound of the small bag, which is 39¢ per pound. Here's your solution with all units attached, division treated as it should be, and rounded to the hundredth-place:

$$50\text{ lb} \div \$11 = \frac{50\text{ lb}}{\$11} = 4.55\frac{\text{lb}}{\$}.$$

This actually produces some useful information, as you can see a dollar buys 4.55 pounds of flour. This is a better buy than 39¢ per pound, which is roughly 3 pounds for a dollar. But it's better to find the price per pound for the large bag:

$$\$11 \div 50\text{ lb} = \frac{\$11}{50\text{ lb}} = 0.22\frac{\$}{\text{lb}} = 22¢\text{ per pound.}$$

Now you can compare this value to the price per pound of the small bag (39¢/lb). The large bag is the better buy.

45. $32\text{ ft} \times \frac{12\text{ in}}{1\text{ ft}} = 384\text{ in}$

47. $35\text{ min} \times \frac{60\text{ s}}{1\text{ min}} = 2100\text{ s}$

49. $4.2\text{ hr} \times \frac{60\text{ min}}{1\text{ hr}} \times \frac{60\text{ s}}{1\text{ min}} = 15{,}120\text{ s}$

51. $4\text{ yr} \times \frac{365\text{ day}}{1\text{ yr}} \times \frac{24\text{ hr}}{1\text{ day}} = 35{,}040\text{ hr}$

53. Note that 1 ft = 12 in, and thus $(1\text{ ft})^2 = (12\text{ in})^2$, which means $1\text{ ft}^2 = 144\text{ in}^2$. This can also be written as $\frac{1\text{ ft}^2}{144\text{ in}^2} = 1$, or $\frac{144\text{ in}^2}{1\text{ ft}^2} = 1$.

55. $3.5\text{ acres} \times \frac{43{,}560\text{ ft}^2}{1\text{ acre}} = 152{,}460\text{ ft}^2$

57. Since 1 m = 100 cm, we have $(1\text{ m})^3 = (100\text{ cm})^3$, which means $1\text{ m}^3 = 1{,}000{,}000\text{ cm}^3$. This can also be written as $\frac{1\text{ m}^3}{1{,}000{,}000\text{ cm}^3} = 1$, or $\frac{1{,}000{,}000\text{ cm}^3}{1\text{ m}^3} = 1$.

59. Since 1 yd = 3 ft, we have $(1\text{ yd})^3 = (3\text{ ft})^3$, or $1\text{ yd}^3 = 27\text{ ft}^3$. Therefore $350\text{ ft}^3 \times \frac{1\text{ yd}^3}{27\text{ ft}^3} = 13.0\text{ yd}^3$.

61. $82\text{ pounds} \times \frac{\$1.221}{1\text{ pound}} = \$100.12$ or $82\text{ Pounds} \times \frac{\$1}{0.8191\text{ Pound}} = \$100.11$

63. $320\text{ euros} \times \frac{\$1.058}{1\text{ euro}} = \$338.56$ or $320\text{ euros} \times \frac{\$1}{0.9449\text{ euro}} = \$338.66$

65. $\frac{1.5\text{ euro}}{1\text{ L}} \times \frac{3.785\text{ L}}{1\text{ gal}} \times \frac{\$1.058}{1\text{ euro}} = \frac{\$6.01}{\text{gal}}$

FURTHER APPLICATIONS

67. The train is traveling with speed $\frac{45\text{ mi}}{34\text{ min}}$, and this needs to be converted to mi/hr.

$$\frac{45\text{ mi}}{34\text{ min}} \times \frac{60\text{ min}}{1\text{ hr}} = 79.4\ \frac{\text{mi}}{\text{hr}}$$

69. Convert 3,998,000 births per year into births per minute. $\frac{3{,}998{,}000\text{ births}}{1\text{ yr}} \times \frac{1\text{ yr}}{365\text{ d}} \times \frac{1\text{ d}}{24\text{ hr}} \times \frac{1\text{ hr}}{60\text{ min}}$ $= 7.6\frac{\text{births}}{\text{min}}$. Note that 321 million people is 321,000 groups of 1000 people each. The annual birth rate is $\frac{3{,}998{,}000\text{ births}}{321{,}000{,}000\text{ people}} \times \frac{321{,}000{,}000\text{ people}}{321{,}000\text{ groups of }1000} = 12.5\text{ births/100,000 people.}$

71. Convert one lifetime into heart beats. $1\text{ lifetime} \times \frac{80\text{ yr}}{\text{lifetime}} \times \frac{365\text{ d}}{\text{yr}} \times \frac{24\text{ hr}}{1\text{ d}} \times \frac{60\text{ min}}{1\text{ hr}} \times \frac{65\text{ beats}}{1\text{ min}}$ $= 2{,}733{,}120{,}000\text{ beats}$

73. a. One full tank of gas for Car A costs $12\text{ gal} \times \frac{\$2.55}{1\text{ gal}} = \$30.60$ and one full tank of gas for Car B costs $20\text{ gal} \times \frac{\$2.55}{1\text{ gal}} = \$51.00.$

b. Car A will use $3000\text{ mi} \times \frac{1\text{ gal}}{40\text{ mi}} \times \frac{1\text{ tank}}{12\text{ gal}} = 6.25$ tanks of gas and Car B will use $3000\text{ mi} \times \frac{1\text{ gal}}{30\text{ mi}} \times \frac{1\text{ tank}}{20\text{ gal}}$ $= 5$ tanks of gas.

c. The driver of Car A will spend $6.25\text{ tanks} \times \frac{\$30.60}{1\text{ tank}} = \$191.50$ and the driver of Car B will spend $5\text{ tanks} \times \frac{\$51.00}{1\text{ tank}} = \$255.00.$

75. a. The driving time when traveling at 60 miles per hour is $1500\text{ mi} \times \frac{1\text{ hr}}{60\text{ mi}} = 25\text{ hr},$ while the time at 75 miles per hour is $1500\text{ mi} \times \frac{1\text{ hr}}{75\text{ mi}} = 20\text{ hr}.$

b. Your car gets 32 miles to the gallon when driving at 60 mph, so the cost is: $1500\text{ mi} \times \frac{1\text{ gal}}{32\text{ mi}} \times \frac{\$2.55}{\text{gal}}$ $= \$119.53.$

Your car gets 25 miles to the gallon when driving at 75 mph, so the cost is: $1500\text{ mi} \times \frac{1\text{ gal}}{25\text{ mi}} \times \frac{\$2.55}{\text{gal}}$ $= \$153.00.$

77. Convert 6 breaths per minute into liters of warmed air per day using 1 minute = 6 breaths, and 1 breath = 0.5 L.

$$\frac{6\text{ breaths}}{1\text{ min}} \times \frac{60\text{ min}}{1\text{ hr}} \times \frac{0.5\text{ L}}{1\text{ breath}} \times \frac{24\text{ hr}}{1\text{ day}} = 4320\frac{\text{L}}{\text{day}}$$

79. Answers will vary depending on prices. The area of the yard is $60\text{ ft} \times 35\text{ ft} = 2100\text{ ft}^2$.

a. The cost to plant the region with grass seed would be $\frac{\text{Cost of grass seed}}{1\text{ ft}^2} \times 2100\text{ ft}^2$.

79. (continued)

b. The cost to cover the region with sod would be $\frac{\text{Cost of sod}}{1 \text{ ft}^2} \times 2100 \text{ ft}^2$.

c. The cost to cover the region with topsoil and plant bulbs would be $\frac{\text{Cost of topsoil}}{\text{ft}^2} \times 2100 \text{ ft}^2$

$+ \frac{\text{Cost}}{1 \text{ bulb}} \times \frac{2 \text{ bulbs}}{1 \text{ ft}^2} \times 2100 \text{ ft}^2$.

USING TECHNOLOGY

85. Answers will vary.

UNIT 2B: EXTENDING UNIT ANALYSIS

QUICK QUIZ

1. **a.** Divide both sides of $1 \text{ L} = 1.057 \text{ qt}$ by 1 L.
2. **a.** Since one meter is a little longer than one yard, 1200 meters will be longer than 3600 feet.
3. **b.** Intuitively, it makes sense to divide a volume (with units of, say, ft^3) by a depth (with units of ft) to arrive at surface area (which has units of ft^2).
4. **b.** A watt is defined to be one joule per second, which is a unit of power, not energy.
5. **c.** Energy used by an appliance is computed by multiplying the power rating by the amount of time the appliance is used, so you need to know how long the light bulb is on in order to compute the energy it uses in that time span.
6. **a.** Multiplying the population density, which has units of $\frac{\text{people}}{\text{mi}^2}$, by the area, which has units of mi^2, will result in an answer with units of people.
7. **c.** Water boils at 100°C (at sea level), so 110°C is boiling hot.
8. **b.** Concentrations of gases are often stated in parts per million (and the other two answers make no sense).
9. **c.** Multiplying by 30 kg will give the daily dose, which should be divided by three to determine the dose for one eight hour period.
10. **b.** $4 \text{ L} \times \frac{0.08 \text{ gm}}{100 \text{ mL}} \times \frac{1000 \text{ mL}}{1 \text{ L}} = 0.08 \text{ gm} \times 40 = 3.2 \text{ gm}$

REVIEW QUESTIONS

1. Standardized units have the same meaning for everyone. Standardized units allow measurements to be used by people with no misunderstanding of what they mean or represent.
3. Energy is what makes matter move or heat up. The international metric unit of energy is the joule. Calories are used to measure the energy our bodies can draw from food, where 1 Calorie = 4184 joules. Electrical energy is usually measured in kilowatt-hours, where 1 kilowatt-hour = 3.6 million joules.
5. Power is the rate at which energy is used. The international metric unit of power is the watt, defined as $1 \text{ watt} = 1 \frac{\text{joule}}{\text{s}}$.

DOES IT MAKE SENSE?

7. Makes sense. Liquids are measured in liters, and since one liter is about a quart, drinking two liters is a reasonable thing to do.
9. Does not make sense. The unit of meter measures length, not volume.

11. Does not make sense. The volume of a sphere is $V = \frac{4}{3}\pi r^3$, so the volume of a beach ball with a radius of 20 cm (about 8 inches) would be more than 32,000 cm^3. This translates into a mass of 320,000 grams if the density is 10 grams per cm^3, which is 320 kg. The mass of a beach ball isn't anywhere near that large.

BASIC SKILLS AND CONCEPTS

13. a. $10^4 \times 10^7 = 10^{4+7} = 10^{11}$

b. $10^5 \times 10^{-3} = 10^{5-3} = 10^2$

c. $10^6 \div 10^2 = 10^{6-2} = 10^4$

d. $\frac{10^8}{10^{-4}} = 10^{8-(-4)} = 10^{12}$

e. $\frac{10^{12}}{10^{-4}} = 10^{12-(-4)} = 10^{16}$

f. $10^{23} \times 10^{-23} = 10^{23-23} = 10^0 = 1$

g. $10^4 + 10^2 = 10{,}000 + 100 = 10{,}100$

h. $10^{15} \div 10^{-5} = 10^{15-(-5)} = 10^{20}$

15. a. $10 \text{ furlongs} \times \frac{1 \text{ mi}}{8 \text{ furlongs}} = 1.25 \text{ miles}$

b. $10 \text{ furlongs} \times \frac{1 \text{ mi}}{8 \text{ furlongs}} \times \frac{5280 \text{ ft}}{1 \text{ mi}} \times \frac{1 \text{ yd}}{3 \text{ ft}} = 2200 \text{ yards}$

17. Convert a cubic foot of water into pounds: $1 \text{ ft}^3 \text{ (of water)} \times \frac{7.48 \text{ gal}}{1 \text{ ft}^3} \times \frac{8.33 \text{ lb}}{1 \text{ gal}} = 62.3 \text{ lb.}$

Now convert that answer into ounces: $62.3 \text{ lb} \times \frac{16 \text{ oz}}{1 \text{ lb}} = 997 \text{ oz (av)}.$

19. a. $30 \text{ knots} = \frac{30 \text{ naut. mi}}{\text{hr}} \times \frac{6076.1 \text{ ft}}{1 \text{ naut. mi}} \times \frac{1 \text{ mi}}{5280 \text{ ft}} = 34.5 \frac{\text{mi}}{\text{hr}}$

b. $102{,}000 \text{ tons} \times \frac{2000 \text{ lb}}{1 \text{ ton}} \times \frac{1 \text{ kg}}{2.205 \text{ lb}} \times \frac{1 \text{ metric tonne}}{1000 \text{ kg}} = 92{,}517 \text{ metric tonnes}$

21. 6-ounce bottle: $\frac{\$3.99}{6 \text{ oz}} = \frac{\$0.665}{1 \text{ oz}}$

14-ounce bottle: $\frac{\$9.49}{14 \text{ oz}} \approx \frac{\$0.678}{1 \text{ oz}}$

The 6-ounce bottle is the better deal.

23. The 15-gallon tank will cost $\frac{\$35.25}{15 \text{ gal}} = \frac{\$2.35}{1 \text{ gal}}$ so the 15-gallon tank option is the better deal.

25. You can tell the factor by which the first unit is larger than the second by dividing the second into the first. Rewrite metric prefixes as powers of 10, and then simplify, as shown below.

$\frac{1 \text{ km}}{1 \text{ m}} = \frac{10^3 \text{ m}}{10^0 \text{ m}} = \frac{10^3}{10^0} = 10^{3-0} = 10^3$. This means a kilometer is 1000 times as large as a meter.

27. You can tell the factor by which the first unit is larger than the second by dividing the second into the first. Rewrite metric prefixes as powers of 10, and then simplify, as shown below.

$\frac{1 \text{ L}}{1 \mu\text{L}} = \frac{10^0 \text{ L}}{10^{-6} \text{ L}} = 10^{0-(-6)} = 10^6$, so a liter is 1,000,000 times larger than a microliter.

29. You can tell the factor by which the first unit is larger than the second by dividing the second into the first. Rewrite metric prefixes as powers of 10, and then simplify, as shown below.

$\frac{1\text{ m}^2}{1\text{ mm}^2} = \frac{10^0\text{ m}^2}{(10^{-3}\text{ m})^2} = \frac{10^0\text{ m}^2}{10^{-6}\text{ m}^2} = 10^{0-(-6)} = 10^6$, so a square meter is 1,000,000 times as large as a square millimeter.

31. $13\text{ L} \times \frac{1.057\text{ qt}}{1\text{ L}} = 13.7\text{ qt}$

33. $34\text{ lb} \times \frac{1\text{ kg}}{2.205\text{ lb}} = 15.4\text{ kg}$

35. Square both sides of the conversion $1\text{ km} = 0.6214\text{ mi}$ to find the conversion between square miles and square kilometers: $(1\text{ km})^2 = (0.6214\text{ mi})^2 \Rightarrow 1\text{ km}^2 = 0.38614\text{ mi}^2$.

Now use this to complete the problem: $3\text{ km}^2 \times \frac{0.38614\text{ mi}^2}{1\text{ km}^2} = 1.2\text{ mi}^2$.

37. $47\frac{\text{mi}}{\text{hr}} \times \frac{1.6093\text{ km}}{1\text{ mi}} \times \frac{1000\text{ m}}{1\text{ km}} \times \frac{1\text{ hr}}{60\text{ min}} \times \frac{1\text{ min}}{60\text{ s}} = 21.0\frac{\text{m}}{\text{s}}$

39. $\frac{23\text{ g}}{1\text{ cm}^3} \times \frac{1\text{ kg}}{1000\text{ g}} \times \frac{2.205\text{ lb}}{1\text{ kg}} \times \left(\frac{2.54\text{ cm}}{1\text{ in}}\right)^3 = 0.8\frac{\text{lb}}{\text{in}^3}$

41. The rise in sea level is found by dividing the volume of water by the Earth's surface area, so

sea level rise $= \frac{2.5\times10^6\text{ km}^3}{340\times10^6\text{ km}^2} = 0.007\text{ km}$ or about 7 meters.

43. a. $C = \frac{45-32}{1.8} = 7.2°\text{C}$

b. $F = 1.8(20)+32 = 68°\text{F}$

c. $F = 1.8(-15)+32 = 5°\text{F}$

d. $F = 1.8(-30)+32 = -22°\text{F}$

e. $C = \frac{70-32}{1.8} = 21.1°\text{C}$

45. a. $C = 50 - 273.15 = -223.15°\text{C}$

b. $C = 240 - 273.15 = -33.15°\text{C}$

c. $K = 10 + 273.15 = 283.15\text{ K}$

47. Power is the rate at which energy is used, so the power is 150 Calories per mile. Convert this to joules per second, or watts. $\frac{150\text{ Cal}}{1\text{ mi}} \times \frac{1\text{ mi}}{6\text{ min}} \times \frac{1\text{ min}}{60\text{ s}} \times \frac{4184\text{ J}}{1\text{ Cal}} = 1743\text{ W}$

49. In order to compute the cost, we need to find the energy used by the bulbs in kW-hr, and then convert to cents. This can be done in a single chain of unit conversions, beginning with the idea that energy = power × time.

100 watt bulb: One day will cost $100\text{ W} \times \frac{12\text{ hr}}{1\text{ day}} \times \frac{1\text{ kW}}{1000\text{ W}} \times \frac{\$0.13}{1\text{ kW-hr}} = \$0.156.$

25 watt bulb: One day will cost $25\text{ W} \times \frac{12\text{ hr}}{1\text{ day}} \times \frac{1\text{ kW}}{1000\text{ W}} \times \frac{\$0.13}{1\text{ kW-hr}} = \$0.039$

You will save $(\$0.156 - \$0.039) \times 365\text{ days} = \$42.71$ in one year.

51. Density is mass per unit volume, so the density of the block is $\frac{0.12\text{ kg}}{200\text{ cm}^3} \times \frac{1000\text{ g}}{1\text{ kg}} = 0.6\frac{\text{g}}{\text{cm}^3}$. It will float in water because the density of water is $1\frac{\text{g}}{\text{cm}^3}$.

53. County A has more people.

County A: $100 \text{ mi}^2 \times \left(\frac{1.6093 \text{ km}}{1 \text{ mi}}\right)^2 \times \frac{25 \text{ people}}{1 \text{ km}^2} = 6474 \text{ people}$

County B: $25 \text{ km}^2 \times \left(\frac{1 \text{ mi}}{1.6093 \text{ km}}\right)^2 \times \frac{100 \text{ people}}{1 \text{ mi}^2} = 965 \text{ people}$

55. The population density for New Jersey is $\frac{9{,}000{,}000 \text{ people}}{7417 \text{ mi}^2} = 1213 \frac{\text{people}}{\text{mi}^2}$. The population density for Alaska is $\frac{738{,}000 \text{ people}}{571{,}951 \text{ mi}^2} = 1.3 \frac{\text{people}}{\text{mi}^2}$, which is smaller than New Jersey's population density.

57. a. In one week, a 100-pound person would take $1 \text{ week} \times \frac{7 \text{ days}}{1 \text{ week}} \times \frac{25 \text{ mg}}{6 \text{ hr}} \times \frac{24 \text{ hr}}{1 \text{ day}} = 700 \text{ mg}$, so should take $700 \text{ mg} \times \frac{1 \text{ tablet}}{12.5 \text{ mg}} = 56$ tablets.

b. A 100-pound person should take $700 \text{ mg} \times \frac{5 \text{ mL}}{12.5 \text{ mg}} = 280 \text{ mL}$ of liquid Benadryl.

59. a. BAC is usually measured in units of grams of alcohol per 100 milliliters of blood. A woman who drinks two glasses of wine, each with 20 grams of alcohol, has consumed 40 grams of alcohol. If she has 4000 milliliters of blood, her BAC is $\frac{40 \text{ g}}{4000 \text{ mL}} = \frac{0.01 \text{ g}}{\text{mL}} \times \frac{100}{100} = \frac{1.0 \text{ g}}{100 \text{ mL}}$. It is fortunate that alcohol is not absorbed immediately, because if it were, the woman would most likely die – a BAC above 0.4 g/mL is typically enough to induce coma or death.

b. If alcohol is eliminated from the body at a rate of 10 grams per hour, then after 3 hours, 30 grams would have been eliminated. This leaves 10 grams in the woman's system, which means her BAC is $\frac{10 \text{ g}}{4000 \text{ mL}} = \frac{0.0025 \text{ g}}{\text{mL}} \times \frac{100}{100} = \frac{0.25 \text{ g}}{100 \text{ mL}}$. This is well above the legal limit for driving, so it is not safe to drive. Of course this solution assumes the woman survives 3 hours of lethal levels of alcohol in her body, because we have assumed all the alcohol is absorbed immediately. In reality, the situation is somewhat more complicated.

FURTHER APPLICATIONS

61. a. A metric mile is $1500 \text{ m} \times \frac{3.28 \text{ ft}}{1 \text{ m}} = 4920 \text{ ft}$. A USCS mile is 5280 ft. Since $4920/5280 = 0.932$, the metric mile is 93.2% of the USCS mile.

b. Men's mile: $\frac{1 \text{ mi}}{223.13 \text{ s}} \times \frac{60 \text{ s}}{1 \text{ min}} \times \frac{60 \text{ min}}{1 \text{ hr}} = 16.13 \text{ mi/hr}$; $(3:43:13 = 223.13 \text{ seconds})$

Men's metric mile: $\frac{4920 \text{ ft}}{206 \text{ s}} \times \frac{1 \text{ mi}}{5280 \text{ ft}} \times \frac{60 \text{ s}}{1 \text{ min}} \times \frac{60 \text{ min}}{1 \text{ hr}} = 16.28 \text{ mi/hr}$; $(3:26:00 = 206 \text{ seconds})$

The record holder in the metric mile ran at a faster pace.

c. Women's mile: $\frac{1 \text{ mi}}{252.56 \text{ s}} \times \frac{60 \text{ s}}{1 \text{ min}} \times \frac{60 \text{ min}}{1 \text{ hr}} = 14.25 \text{ mi/hr}$; $(4:12:56 = 252.56 \text{ seconds})$

Women's metric mile: $\frac{4920 \text{ ft}}{230.07 \text{ s}} \times \frac{1 \text{ mi}}{5280 \text{ ft}} \times \frac{60 \text{ s}}{1 \text{ min}} \times \frac{60 \text{ min}}{1 \text{ hr}} = 14.58 \text{ mi/hr}$; $(3:50:7 = 230.07 \text{ seconds})$

The record holder in the metric mile ran at a faster pace.

d. This would be true in both cases since, for both men and women, the record holder in the metric mile ran at a faster pace.

63. $\frac{1}{2}\text{gal}\times\frac{3.785\text{ L}}{1\text{ gal}}\times\frac{100\text{ pesos}}{0.8\text{ L}}\times\frac{\$1}{21.86\text{ pesos}}=\$10.82$

65. Monte Carlo costs $\frac{1150\text{ euro}}{80\text{ m}^2}\times\left(\frac{1\text{ m}}{3.28\text{ ft}}\right)^2\times\frac{\$1.058}{1\text{ euro}}=\$1.41/\text{ft}^2$ and Santa Fe costs $\frac{\$800}{500\text{ ft}^2}=\$1.60/\text{ft}^2$, so Monte Carlo is less expensive.

67 The Hope diamond weighs 45.52 carats, which is 0.32 oz.

$$45.52\text{ carats}\times\frac{0.2\text{ g}}{1\text{ carat}}=9.1\text{ g and }9.1\text{ g}\times\frac{1\text{ kg}}{1000\text{ g}}\times\frac{2.205\text{ lb}}{1\text{ kg}}\times\frac{16\text{ oz}}{1\text{ lb}}=0.32\text{ oz.}$$

69. 2.2 ounces of 16-karat gold is $2.2\text{ oz}\times\frac{16}{24}=1.47\text{ oz}$ of pure gold.

71. The Cullinan diamond weighs 3106 carats.

$$3106\text{ carats}\times\frac{0.2\text{ g}}{1\text{ carat}}\times\frac{1000\text{ mg}}{1\text{ g}}=621{,}200\text{ mg and }3106\text{ carats}\times\frac{0.2\text{ g}}{1\text{ carat}}\times\frac{1\text{ kg}}{1000\text{ g}}\times\frac{2.205\text{lb}}{1\text{ kg}}=1.37\text{lb}$$

73. a. The volume of the bath is $6\text{ ft}\times 3\text{ ft}\times 2.5\text{ ft}=45\text{ ft}^3$. Fill it to the halfway point, and you'll use 22.5 ft^3 of water (half of 45 is 22.5). (Interesting side note: it doesn't matter which of the bathtub's three dimensions you regard as the height – fill it halfway, and it's always 22.5 ft^3 of water). When you take a shower, you use

$$1\text{ shower}\times\frac{10\text{ min}}{\text{shower}}\times\frac{1.75\text{ gal}}{\text{min}}\times\frac{1\text{ ft}^3}{7.5\text{ gal}}=2.33\text{ ft}^3$$

of water, and thus you use considerably more water when taking a bath.

b. Convert 22.5 ft^3 of water (the water used in a bath) into minutes.

$$22.5\text{ ft}^3\times\frac{7.5\text{ gal}}{1\text{ ft}^3}\times\frac{1\text{ min}}{1.75\text{ gal}}=96\text{ min}$$

c. Plug the drain in the bathtub, and mark the depth to which you would normally fill the tub when taking a bath. Take a shower, and note how long your shower lasts. Step out and towel off, but keep the shower running. When the water reaches your mark (you used a crayon, and not a pencil, right?), note the time it took to get there. You now have a sense of how many showers it takes to use the same amount of water as a bath. For example, suppose your shower took 12 minutes, and it takes a full hour (60 minutes) for the water to reach your mark. That would mean every bath uses as much water as five showers.

75. a. Convert 9 billion gallons per day into cfs: $\frac{9{,}000{,}000{,}000\text{ gal}}{\text{d}}\times\frac{1\text{ ft}^3}{7.5\text{ gal}}\times\frac{1\text{ d}}{24\text{ hr}}\times\frac{1\text{ hr}}{60\text{ min}}\times\frac{1\text{ min}}{60\text{ s}}$

$=13{,}889\frac{\text{ft}^3}{\text{s}}$, or 13,889 cfs, which is about 46% of the flow rate of the Colorado River.

b. The volume of water entering the city over the course of a day, in cubic feet, was

$9{,}000{,}000{,}000\text{ gal}\times\frac{1\text{ ft}^3}{7.5\text{ gal}}=1{,}200{,}000{,}000\text{ ft}^3$.

The area of the flooded portion of the city, in square feet, was $6\text{ mi}^2\times\left(\frac{5280\text{ ft}}{1\text{ mi}}\right)^2=167{,}270{,}400\text{ ft}^2$.

If the volume of water is divided by the area it covers, you'll get the average depth of the water (see Exercise 59): average depth $=\frac{\text{volume}}{\text{area}}=\frac{1{,}200{,}000{,}000\text{ ft}^3}{167{,}270{,}400\text{ ft}^2}=7.2\text{ ft.}$

Thus the water level rose about 7 feet in one day, assuming the entire 6 square miles was covered in that day.

77. a. The volume of one hemlock tree would be about $\pi(15\text{ in})^2(120\text{ ft}) = 84{,}823\text{ in}\times\text{in}\times\text{ft}$ and

$$84{,}823\text{ in}\times\text{in}\times\text{ft}\times\frac{1\text{ ft}}{12\text{ in}} = 7069\text{ ft}\times\text{ft}\times\text{in},\text{ or }7069\text{ fbm}.$$

b. One 8 ft 2-by-4 contains about $8\text{ ft}\times 1.5\text{ in}\times 3.5\text{ in}\times\frac{1\text{ ft}}{12\text{ in}} = 3.5\text{ ft}\times\text{ft}\times\text{in} = 3.5$ fbm, so you could cut

$\frac{150\text{ fbm}}{3.5\text{ fbm}} = 42.9$, or 43 whole boards. (Round up to next whole board.)

c. One 12 ft 2-by-6 is $12\text{ ft}\times 1.5\text{ in}\times 5.5\text{ in}\times\frac{1\text{ ft}}{12\text{ in}} = 8.25\text{ ft}\times\text{ft}\times\text{in} = 8.25$ fbm, so the project would require

$75\times 8.25\text{ fbm}\approx 619$ fbm.

79. Case A:

Income: $50\text{ acres}\times\frac{60\text{ bushels}}{1\text{ acre}}\times\frac{\$3.50}{1\text{ bushel}} = \$10{,}500$

Cost: $50\text{ acres}\times\frac{100\text{ lb}}{1\text{ acre}}\times\frac{\$0.25}{1\text{ lb}} = \$1250$

Revenue: $\$10{,}500 - \$1250 = \$9250$

Case B:

Income: $50\text{ acres}\times\frac{50\text{ bushels}}{1\text{ acre}}\times\frac{\$4.50}{1\text{ bushel}} = \$11{,}250$

Cost: $50\text{ acres}\times\frac{70\text{ lb}}{1\text{ acre}}\times\frac{\$0.50}{1\text{ lb}} = \$1750$

Revenue: $\$11{,}250 - \$1750 = \$9500$

Case B has the higher revenue.

81. a. Convert kilowatt-hours into joules: $900\text{ kW-hr}\times\frac{3{,}600{,}000\text{ J}}{1\text{ kW-hr}} = 3{,}240{,}000{,}000$ J.

b. May has 31 days, so the average power is $\frac{3{,}240{,}000{,}000\text{ J}}{31\text{ d}}\times\frac{1\text{ d}}{24\text{ hr}}\times\frac{1\text{ hr}}{60\text{ min}}\times\frac{1\text{ min}}{60\text{ s}} = \frac{1210\text{ J}}{\text{s}} = 1210$ W.

One could also begin with 900 kW-hr per 31 days, and convert that into watts, using 1 watt = 1 joule per second.

c. First, convert joules into liters: $3{,}240{,}000{,}000\text{ J}\times\frac{1\text{ L}}{12{,}000{,}000\text{ J}} = 270$ L.

Now convert liters into gallons: $270\text{ L}\times\frac{1\text{ gal}}{3.785\text{ L}} = 71.33$ gal.

Thus it would take 270 liters = 71.33 gallons of oil to provide the energy shown on the bill (assuming all the energy released by the burning oil could be captured and delivered to your home with no loss).

83. a. Your power is the rate at which you use energy, and thus your average power is 2500 Calories per day.

Convert this to watts: $\frac{2500\text{ Cal}}{1\text{ d}}\times\frac{1\text{ d}}{24\text{ hr}}\times\frac{1\text{ hr}}{60\text{ min}}\times\frac{1\text{ min}}{60\text{ s}}\times\frac{4184\text{ J}}{1\text{ Cal}} = 121\frac{\text{J}}{\text{s}}$. Since $1\text{ W} = 1\text{ J/s}$, this is 121 watts.

b. $\frac{2500\text{ Cal}}{1\text{ d}}\times\frac{365\text{ d}}{1\text{ yr}}\times\frac{4184\text{ J}}{1\text{ Cal}} = 3{,}817{,}900{,}000\frac{\text{J}}{\text{yr}}$, so you need about 3.8 billion joules each year from food, which is very close to 1% of your total energy consumption (3.8 billion/400 billion = 0.0095).

85. Since energy = power × time, the power plant can produce (assuming a 30-day month) 1190 MW × 1 month $\times\frac{30\text{ d}}{1\text{ month}}\times\frac{24\text{ hr}}{1\text{ d}}\times\frac{1000\text{ kW}}{1\text{ MW}}\approx 8.57\times10^{8}$ kW-hr of energy each month. The amount of uranium used each month is given by $8.57\times10^{8}\ \frac{\text{kW-hr}}{\text{month}}\times\frac{1\text{ kg}}{16\times10^{6}\text{ kW-hr}}\approx 53.6\ \frac{\text{kg}}{\text{month}}$. The plant can serve about 857,000 homes, because $8.57\times10^{8}\text{ kW-hr}\times\frac{1\text{ home}}{1000\text{ kW-hr}}=857{,}000$ homes.

87. At 20% efficiency, this solar panel can generate 200 watts of power when exposed to direct sunlight. Since energy = power × time, and because the panel receives the equivalent of 6 hours of direct sunlight, the panel can produce $200\text{ W}\times 6\text{ hr}\times\frac{1\text{ kW}}{1000\text{ W}}\times\frac{3{,}600{,}000\text{ j}}{\text{kW-hr}}=4{,}320{,}000\text{ j}$. This occurs over the span on one day, so the panel produces an average power of $\frac{4{,}320{,}000\text{ j}}{\text{d}}\times\frac{1\text{ d}}{24\text{ hr}}\times\frac{1\text{ hr}}{60\text{ min}}\times\frac{1\text{ min}}{60\text{ s}}=50\frac{\text{j}}{\text{s}}=50\text{ W}$.

89. Energy = power × time, so the energy produced by a wind turbine over the course of a year is $2.5\text{ MW}\times\frac{1000\text{ kW}}{1\text{ MW}}\times 1\text{ yr}\times\frac{365\text{ d}}{1\text{ yr}}\times\frac{24\text{ hr}}{1\text{ d}}=21{,}900{,}000\text{ kW-hr}$. This is enough energy to serve $21{,}900{,}000\text{ kW-hr}\times\frac{1\text{ household}}{10{,}000\text{ kW-hr}}=2190$ households.

91. a. There are $750\text{ mL}\times\frac{5\text{ mg}}{100\text{ mL}}=37.5\text{ mg}$ of dextrose in 750 mL of D5W. The patient should be given $50\text{ mg}\times\frac{750\text{ mL}}{37.5\text{ mg}}=1000\text{ mL}$ of D5W.

b. There are $1.2\text{ L}\times\frac{1000\text{ mL}}{1\text{ L}}\times\frac{0.9\text{ mg}}{100\text{ mL}}=10.8\text{ mg}$ of dextrose in 1.2 L of NS. The patient should be given $15\text{ mg}\times\frac{1.2\text{ L}}{10.8\text{ mg}}\times\frac{1000\text{ mL}}{1\text{ L}}=1667\text{ mL}$ of NS.

93. a. The flow rates are $\frac{0.5\text{ L}}{4\text{ hr}}\times\frac{1000\text{ mL}}{1\text{ L}}=125\frac{\text{mL}}{\text{hr}}$ and $125\frac{\text{mL}}{\text{hr}}\times\frac{0.9\text{ mg}}{100\text{ L}}=1.125\frac{\text{mg}}{\text{hr}}$.

b. The rate is $125\frac{\text{mL}}{\text{hr}}\times\frac{20\text{ gtt}}{1\text{ mL}}=2500\frac{\text{gtt}}{\text{hr}}$.

c. $4\text{ hr}\times 1.125\frac{\text{mg}}{\text{hr}}=4.5\text{ mg}$ of sodium chloride is delivered.

95. a. $40\text{ kg}\times\frac{25\text{ mg/kg}}{1\text{ day}}=1000\text{ mg/day}$ and $\frac{1000\text{ mg}}{1\text{ day}}\times\frac{1\text{ capsule}}{250\text{ mg}}=4$ capsules/day, so the patient should take 1 capsule every six hours.

b. From part (a), you would need 250 mg in 6 hours, which requires 5 mL of solution in 6 hours, so the rate would be $\frac{5\text{ mL}}{6\text{ hr}}\times\frac{60\text{ gtt}}{1\text{ mL}}=50\text{ gtt/hr}$.

97 a. $0.5\text{ L}\times\frac{20\text{ gtt}}{1\text{ mL}}\times\frac{1000\text{ mL}}{1\text{ L}}=10{,}000\text{ gtt}$

b. $1\text{ L}\times\frac{60\text{ gtt}}{1\text{ mL}}\times\frac{1000\text{ mL}}{1\text{ L}}=60{,}000\text{ gtt}$

c. $1\text{ L}\times\frac{15\text{ gtt}}{1\text{ mL}}\times\frac{1000\text{ mL}}{1\text{ L}}=15{,}000\text{ gtt}$; The rate of infusion is $\frac{15{,}000\text{ gtt}}{5\text{ hr}}\times\frac{1\text{ hr}}{60\text{ min}}=50\text{ gtt/min.}$

UNIT 2C: PROBLEM-SOLVING HINTS

QUICK QUIZ

1. **c.** Look at example 1 (*Box Office Receipts*) in this unit.
2. **a.** Many mathematical problems have more than one solution.
3. **b.** Common experience tells us that batteries can power a flashlight for many hours, not just a few minutes nor several years.
4. **b.** An elevator that carried only 10 kg couldn't accommodate even one person (a 150 lb person weighs about 75 kg), and hotel elevators aren't designed to carry hundreds of people (you'd need at least 100 people to reach 10,000 kg).
5. **b.** Refer to the discussion of Zeno's paradox in the text. There, it was shown that the sum of an infinite number of ever-smaller fraction is equal to 2, and thus we can eliminate answers **a** and **c.** This leaves **b.**
6. **a.** If you cut the cylinder along its length, and lay it flat, it will form a rectangle with width equal to the circumference of the cylinder (i.e. 10 in), and with length equal to the length of the cylinder.
7. **c.** Solving a simpler problem will many times guide you to the solution to a more complex problem.
8. **c.** It could happen that the first 20 balls selected are odd, in which case the next two would have to be even, and this is the first time one can be certain of selecting two even balls.
9. **b.** The most likely explanation is that the A train always arrives 10 minutes after the B train. Suppose the A train arrives on the hour (12:00, 1:00, 2:00,…), while the B train arrives ten minutes before the hour (11:50, 12:50, 1:50,…). If Karen gets to the station in the first 50 minutes of the hour, she'll take the B train; otherwise she'll take the A train. Since an hour is 60 minutes long, 5/6 of the time, Karen will take the B train to the beach. Note that 5/6 of 30 days is 25 days, which is the number of times Karen went to the beach. The other scenarios *could* happen, but they aren't nearly as likely as the scenario in answer **b.**
10. **b.** Label the hamburgers as A, B, and C. Put burgers A and B on the grill. After 5 minutes, turn burger A, take burger B and put it on a plate, and put burger C on the grill. After 10 minutes, burger A is cooked, while burgers B and C are half-cooked. Finish off burgers B and C in the final 5 minutes, and you've cooked all three in 15 minutes.

REVIEW QUESTIONS

1. Examples will vary.

   Hint 1: There May Be More Than One Answer: Some problems may have more than one mathematically correct solution. Multiple solutions often occur because not enough information is available to distinguish among a variety of possibilities.

   Hint 2: There May Be More Than One Method: Although there may be more than one method for finding an answer, not all methods are equally efficient. An efficient method can save a lot of time and work.

   Hint 3: Use Appropriate Tools: You usually will have a choice of tools to use in any problem. Choosing the tools most suited to the job will make your task much easier.

   Hint 4: Consider Simpler, Similar Problems: Solving a simpler, but similar, problem may provide insight to help you understand the original problem.

1. (continued)

Hint 5: Consider Equivalent Problems with Simpler Solutions: An equivalent problem will have the same numerical answer but may be easier to solve.

Hint 6: Approximations Can Be Useful: Approximations may reveal the essential character of a problem, making it easier to reach an exact solution. Approximations also provide a useful check: If you come up with an "exact solution" that isn't close to the approximate one, something may have gone wrong.

Hint 7: Try Alternative Patterns of Thought: Approach every problem with an openness that allows innovative ideas to percolate.

Hint 8: Do Not Spin Your Wheels: Often the best strategy in problem solving is to put a problem aside for a few hours or days.

DOES IT MAKE SENSE?

3. Does not make sense. There is no problem-solving recipe that can be applied to all problems.

5. Does not make sense. Approximations can used to simplify calculations and can result in solutions to problems that are accurate enough for real world use.

BASIC SKILLS AND CONCEPTS

7. You won't be able to determine the exact number of cars and buses that passed through the toll booth, but the method of trial-and-error leads to the following possible solutions: (16 cars, 0 buses), (13 cars, 2 buses), (10 cars, 4 buses), (7 cars, 6 buses), (4 cars, 8 buses), (1 car, 10 buses). Along the way, you may have noticed that the number of buses must be an even number, because when it is odd, the remaining money cannot be divided evenly into $2 (car) tolls.

9. a. Based on the first race data, Jordan runs 200 meters in the time it takes Amari to run 190 meters. In the second race, Jordan will catch up to Amari 10 meters from the finish line (because Jordan has covered 200 meters at that point, and Amari has covered 190 meters). Jordan will win the race in the last ten meters, because Jordan runs faster than Amari.

b. Jordan will be 5 meters from the finish line when Amari is 10 meters from the finish line (because Jordan has covered 200 meters at that point, and Amari has covered 190 meters). Jordan will win the race in the because Jordan runs faster than Amari and is in the lead.

c. Jordan will be 15 meters from the finish line when Amari is 10 meters from the finish line (because Jordan has covered 200 meters at that point, and Amari has covered 190 meters). In the time Amari runs 10 meters, Jordan will only run $10\ \text{m}_{\text{Amari}} \times \dfrac{200\ \text{m}_{\text{Jordan}}}{190\ \text{m}_{\text{Amari}}} = 10.53\ \text{m}_{\text{Jordan}}$, so Amari wins the race.

d. From part (c), Jordan must start 10.53 meters behind the starting line, since that is how far she would run in the same time Amari finishes the last 10 meters of the race.

11. On the second transfer, there are four possibilities to consider. A) All three marbles are black. B) Two are black, one is white. C) One is black, two are white. D) All three are white. In case A), after the transfer, there are no black marbles in the white pile, and no white marbles in the black pile. In case B), one white marble is transferred to the black pile, and one black marble is left in the white pile. In case C), two white marbles are transferred to the black pile, and two black marbles are left in the white pile. In case D), three white marbles are transferred to the black pile, and three black marbles are left in the white pile. In all four cases, there are as many white marbles in the black pile as there are black marbles in the white pile after the second transfer.

13. Proceeding as in example 6, the circular arc formed by the bowed track can be approximated by two congruent right triangles. Since 1 km is 100,000 cm, the base of one of the triangles is 50,000 cm, and its hypotenuse is 50,005 cm. The height of the triangle can be computed as $h = \sqrt{50{,}005^2 - 50{,}000^2} = 707$ cm, which is about 7.1 meters.

15. Yes, imagine that at the same time the monk leaves the monastery to walk up the mountain, his twin brother leaves the temple and walks down the mountain. Clearly, the two must pass each other somewhere along the path.

FURTHER APPLICATIONS

17. a. Yes; $P=2(7)+2(3)=20$; $A=7\times3=21$

    b. Yes; $P=2(8)+2(2)=20$; $A=8\times2=16$

    c. By trial and error, possible dimensions (l,w) are:

    $(9,1)$; $P=2(9)+2(1)=20$; $A=9\times1=9$

    $(6,4)$; $P=2(6)+2(4)=20$; $A=6\times4=24$

    $(5,5)$; $P=2(5)+2(5)=20$; $A=5\times5=25$

    Note that reversing the length and width will result in the same areas, so the maximum area is when the yard is a square with sides of length 5 meters.

19. Following the hint given in the problem, the figure below shows the room cut along its vertical corners and laid flat with the ceiling in the center. The shortest route for the ant is shown as the solid line, which is the hypotenuse of a right triangle with sides shown as dashed lines. The length of the hypotenuse (using the Pythagorean Theorem) is $\sqrt{(12\text{ ft}-1\text{ ft}-1\text{ ft})^2+(7\text{ ft}+10\text{ ft}+1\text{ ft})^2}=\sqrt{(10\text{ ft})^2+(18\text{ ft})^2}=\sqrt{424\text{ ft}^2}\approx20.6\text{ ft}$.

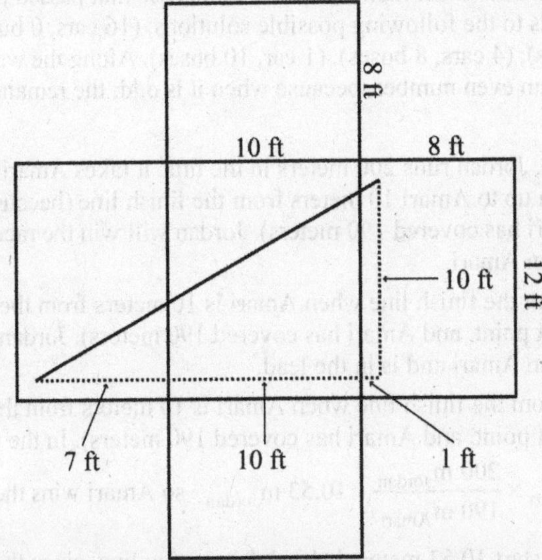

21. In going from the first floor to the third floor, you have gone up two floors. Since it takes 30 seconds to do that, your pace is 15 seconds per floor. Walking from the first floor to the sixth floor means walking up five floors, and at 15 seconds per floor, this will take 75 seconds.

23. You might pick one of each kind of apple on the first three draws. On the fourth draw, you are guaranteed to pick an apple that matches one of the first three, so four draws are required.

25. Select a fruit from the box labeled *Apples and Oranges*. If it's an apple, that box must be the *Apple* box (because its original label is incorrect, leaving only *Apples* or *Oranges* as its correct label). The correct label for the box labeled *Oranges* is either *Apples* or *Apples and Oranges*, but we just determined the box containing apples, so the correct label for this box is *Apples and Oranges*. There's only one choice left for the box labeled *Apples*, and that's *Oranges*. A similar argument allows one to determine the correct labeling for each box if an orange is selected first.

27. Seven crossings are required. Trip 1: the woman crosses with the goose. Trip 2: the woman returns to the other shore by herself. Trip 3: the woman takes the mouse across. Trip 4: the woman returns with the goose. Trip 5: the woman brings the wolf across. Trip 6: the woman returns to the other shore by herself. Trip 7: the woman brings the goose over. In this way, the goose is never left alone with the wolf, nor the mouse with the goose. The foursome could get to the other side of the river in five crossings if you give rowing abilities to the animals, but the statement of the problem (and common sense) probably forbids it (the boat "…will hold only herself and one other animal"). Try it anyway for an interesting twist to the problem.

29. It's possible you'll pick a sock of each color on the first two draws, but on the third draw, you are assured of making a match.

31. When a clock chimes five times, there are four pauses between the chimes. Since it takes five seconds to chime five times, each pause lasts 1.25 seconds. There are nine pauses that need to be accounted for when the clock chimes 10 times, and thus it takes $9 \times 1.25 = 11.25$ seconds to chime 10 times. This solution assumes it takes no time for the clock to chime once (we'd need to know the duration of the chime to solve it otherwise).

33. – 39. Answers will vary.

UNIT 3A: USES AND ABUSES OF PERCENTAGES

QUICK QUIZ

1. **b.** A quantity triples when it is increased by 200%.

2. **b.** The absolute change is $75,000 - 50,000 = 25,000$ and the relative change is $\frac{75,000-50,000}{50,000} = 0.5 = 50\%$.

3. **c.** A negative relative change corresponds to a decreasing quantity.

4. **c.** Suppose Joshua scored 1000 on the SAT. Then Emily would have scored 1500 (50% more than 1000), and Joshua's score is clearly two-thirds of Emily's score (1000/1500 = 2/3).

5. **c.** 120% of \$10 is \$12.

6. **c.** Multiplying a pre-tax price by 1.09 is equivalent to adding 9% tax to the price, and thus dividing a post-tax price by 1.09 will result in the original price. To find out how much tax was paid, take the post-tax price, and subtract the original price: $\$47.96 - \frac{\$47.96}{1.09}$.

7. **a.** The relative change in interest rates that go from 4% to 6% is $\frac{6\% - 4\%}{4\%} = 0.5 = 50\%$.

8. **c.** When a price is decreased by 100%, it drops to zero, as the entire amount of the price is deducted from the original price.

9. **c.** 10% of \$1000 is \$100, so your monthly earnings after one year will be \$1100. Each year thereafter, your earnings will increase by more than \$100 per year (because 10% of \$1100 is more than \$100), and the end result will be more than \$1500.

10. **b.** The only thing we can conclude with certainty is that she won between 20% and 30% of her races. To prove this rigorously, one could detour into an algebra calculation (let $x =$ number of races entered in high school, let $y =$ number of races entered in college, and show that $\frac{0.3x + 0.2y}{x+y}$ is between 20% and 30%). Try different values for x and y, and compute the win percentage of all the races Emily entered to see that answers **a** and **c** are not correct.

REVIEW QUESTIONS

1. Percentages can be used to represent fractions with a denominator of 100. For example, 50% of 1000 is $50\% \times 1000 = (50/100) \times 1000 = 500$. Fractions can also be used to represent change, as in "The increase of 500 people represents a 50% increase in the population." Percentages can also be used for comparisons, as in "The new cost of the jacket is 20% more than the previous price."

3. Absolute differences describe the actual increase or decrease from a reference value to a new value. Relative difference is the size of the absolute difference in comparison to the reference value and can be expressed as a percentage. For example, the absolute difference in the price of a jacket might be $\$120 - \$100 = \$20$, which represents a relative difference of $\frac{\$120 - \$100}{\$100} \times 100\% = 20\%$.

5. When you see a change or difference expressed in percentage points, you can assume it is an absolute change or difference. If it is expressed with the % sign or the word percent, it should be a relative change or difference.

DOES IT MAKE SENSE?

7. Makes sense. If a country's population is declining, then the percent change will be negative (and several of Europe's countries have, indeed, experienced recent declines in population).

9. Does not make sense. You cannot decrease your caloric intake by more than 100% (and you wouldn't want to do that for too long).

11. Does not make sense. Increasing a quantity by several successive percent changes is not equivalent to a one-time percent decrease that is the sum of the successive changes. This can be seen with a simple example: take \$100 and increase it by 10% to \$110, followed by another 10% increase to \$121. Now take the same \$100 and increase it by 20% to \$120. Two successive 10% changes is not equivalent to a single 20% change, just like ten successive 10% changes is not equivalent to a 100% change.

13. Makes sense. If the rate of return on the fund used to be 10%, then a 50% increase would result in a rate of return of 15%.

BASIC SKILLS AND CONCEPTS

15. $2/5 = 0.4 = 40\%$

17. $0.20 = 1/5 = 20\%$

19. $150\% = 1.5 = 3/2$

21. $4/9 = 0.444... = 44.44...\%$

23. $5/8 = 0.625 = 62.5\%$

25. $69\% = 0.69 = 69/100$

27. $7/5 = 1.4 = 140\%$

29. $4/3 = 1.333... = 133.33...\%$

31. 52,252 to 12,333 = 4.24; 12,333 to 52,252 = 0.24; 52,252 is 424% of 12,333

33. 1.6 million to 2.1 million = 0.76; 2.1 million to 1.6 million = 1.31; 1.6 million is 76 % of 2.1 million.

35. 69.50 million to 62.98 million = 1.10; 62.98 million to 69.50 million = 0.91; 69.50 million is 110% of 62.98 million.

37. 302 is 56.1% of 538 because $302/538 = 0.561$.

39. 50,383 is 127% of 39,621 because $50{,}383/39{,}621 = 1.27$.

41. 39.1 million is 115% of 33.9 million because $39.1/33.9 = 0.115$.

43. Clint's absolute change in salary was $\$35{,}000 - \$25{,}000 = \$10{,}000$, and the relative change was $\frac{\$35{,}000 - \$25{,}000}{\$25{,}000} = 0.4 = 40\%$. Helen's absolute change was $\$42{,}000 - \$30{,}000 = \$12{,}000$, and the relative change was $\frac{\$42{,}000 - \$30{,}000}{\$30{,}000} = 0.4 = 40\%$. Thus Helen's salary grew more in absolute terms, but the relative change was the same. This happened because we are comparing Helen's larger absolute change to her also larger initial salary.

45. The absolute change is $16.1 - 8.7 = 7.4$ million. The relative change is $\frac{16.1 - 8.7}{8.7} = 0.851 = 85.1\%$.

47. The absolute change is $55.4 - 67.8 = -12.4$ pounds. The relative change is $\frac{55.4 - 67.8}{67.8} = -0.183 = -18.3\%$.

49. The relative difference is $\frac{81.2 - 76.4}{76.4} = 0.0628 = 6.28\%$. Which means the life expectancy (at birth) of American women (81.2 years) is 6.28% percent greater than the life expectancy of men (76.4 years).

51. The relative difference is $\frac{29.2 - 27.1}{27.1} = 0.0775 = 7.75\%$. Which means in 2015, the median age at first marriage of U.S. women (27.1 years) was 7.75% percent less than the median age at first marriage of U.S. men (29.2 years).

53. Michigan's population is 163% of Missouri's population because 63% more than expresses a relative difference in population, and this is equivalent to $100\% + 63\%$ of the reference population (i.e. Missouri's population).

55. Hawaii's population is 48% of Arkansas's population because 52% *less than* expresses a relative difference in population, and this is equivalent to $100\% - 52\%$ *of* the reference population (i.e. Arkansas's population).

57. Since the wholesale price is 30% *less than* the retail price, it is 70% *of* the retail price, which means it is 0.7 times the retail price.

59. Since the original retail price is 20% *more than* the on-sale price, it is 120% *of* the original retail price, which means it is 1.2 times the on-sale cost.

61. The absolute change is $55.3\% - 40.9\% = 14.4$ percentage point. The relative change is $\frac{55.3-40.9}{40.9}$ $= 0.352 = 35.2\%$.

63. The absolute change is $35.3\% - 50.0\% = -14.7$ percentage points. The relative change is $\frac{35.3-50.0}{50.0}$ $= -0.294 = 29.4\%$.

65. In the first case, 33% of city employees in Freetown ride the bus to work because 10% *more than* 30% is 33%. In the second case, 40% of the city employees in Freetown ride the bus, because 40% is 10 percentage points more than 30%.

67. The pre-tax price is $\frac{\$1278.24}{1.076} = \$1187.96$.

69. The usage rate in 2014 was $\frac{35.3\%}{100\% - 5.6\%} = \frac{0.353}{0.944} = 0.374 = 37.4\%$.

71. False. If the national unemployment rate began at R, and it shrinks at 2% annually, it would be at $R \times 0.98 \times 0.98 \times 0.98 = 0.941R$, which is about 5.9% less than the original value after three years, not 6%.

73. False. Assign an arbitrary number to your profits before the increases, such as \$100. If your profits increase by 11% in the first year, they will be \$111. If they decrease by 3% in the next year, they will be $0.97 \times \$111 = \$107.67$. This is an increase of 7.67%, not 8%.

75. Not possible. A bill can decrease at most by 100%. Once you've deducted 100% from a bill, there's nothing more to pay.

77. Possible. An increase of 100% means hotel rooms cost twice as much now.

79. Possible. This means your computer has three times the storage of my computer.

81. No, the average for the entire course is not 85%. You cannot average averages. As an example, if you have 80% of 400 points in the class going into the final and a 90% on the 100-point final, you have a total of $0.80 \times 400 + 0.90 \times 100 = 410$ points, out of 500 total points. Your class average would be $410/500 = 0.82$, or 82%. (Note: If the final exam is worth half the points in the entire course, your class average would be 85%.)

FURTHER APPLICATIONS

83. True; 20% of 40% is 1/5 of 40%, which is 8%.

85. False. Some of the cars may have both Bluetooth and GPS, and those would be counted twice if the percentages were just added together.

87. The total number of undergraduates is $4550/0.85 = 5353$.

89. *Game Informer Magazine* has $2,400,000/12.5 = 192,000$ digital subscribers.

91. Water usage was reduced by $1 - 0.31 = 0.69$, or 69/100.

93. The relative change is $\frac{8.5-7.1}{7.1} = 0.197 = 19.7\%$.

95. a. Before the loss, the Dow Jones Industrial average was at $10,365 + 778 = 11,143$, so the percentage loss was $\frac{778}{11,143} = 0.0698 = 6.98\%$.

95. (continued)

b. Before the loss, the Dow Jones Industrial average was at $\frac{1739}{0.774} = 2247$, so the point loss is $2247 - 1739 = 508$.

UNIT 3B: PUTTING NUMBERS IN PERSPECTIVE

QUICK QUIZ

1. **a.** The decimal point must move seven places to the left to transform 70,000,000 into scientific notation, and thus its value is 7×10^7.
2. **c.** 1277 is roughly 10^3, and 14,385 is of order 10^4, so their product is around 10^7.
3. **c.** 10^9 is larger than 10^5 by a factor of 10^4, which is ten thousand.
4. **c.** One person, who drives 15,000 miles in a year in a vehicle that gets 25 mpg, uses 600 gallons of gas per year. There are 300 million Americans, and if each used this much gas, the total would be $3 \times 10^8 \text{ people} \times \frac{6 \times 10^2 \text{ gal}}{\text{person}} = 18 \times 10^{10} \text{ gal} = 1.8 \times 10^{11}$ gallons of gasoline, which is 180 billion gallons. Of course, not everyone drives 15,000 miles every year (children, the elderly, and urban dwellers come to mind). On the other hand, this estimate doesn't even consider all the gas used by trucking, airlines, and other industries. Certainly the answer can't be **a** or **b**, as these estimates are much too small. 100 billion gallons is probably a conservative estimate.
5. **c.** A dollar bill is about 6 inches long, or about 15 cm long, and thus 8 million of them amounts to $8 \times 10^6 \text{ bills} \times \frac{1.5 \times 10^1 \text{ cm}}{\text{bill}} \times \frac{1 \text{ km}}{10^5 \text{ cm}} = 12 \times 10^2 \text{ km}$, which is 1200 kilometers. This is enough to get into outer space, but not near enough to reach the moon.
6. **c.** If you know the song from the Broadway musical *Rent*, you may recall that there are 525,600 minutes in a year. 1.2×10^{-10} year is just over one ten-billionth of a year, which isn't even close to a minute (it's a small fraction of a second). Clearly the sun won't burn out in the next fraction of a second, nor in the next minute.
7. **b.** If 1 inch = 100 miles, then 3.5 inches is $3.5 \times 100 = 350$ miles.
8. **b.** \$1 billion is \$1,000 million, so at \$20 million per year, it will take 50 years to earn \$1 billion.
9. **c.** Assuming you're willing to campaign for eight hours per day (which is roughly 500 minutes), and that you'll limit yourself to 10 minutes of talking at each house (including travel time between houses), you'll be able to visit 50 households per day, or 100 houses in two days. This translates into 500 houses in two weeks (campaigning five days per week), or 1000 households in a month, or 12,000 households in a year-long campaign, where you do nothing but campaigning. Clearly it would be impossible to carry out your plan before the election – pay for some TV advertisements instead.
10. **a.** The largest Division 1 college football stadiums hold about 100,000 fans. In a lottery with the stated odds, there's only one winning ticket for every 1,000,000 printed, which means there are 999,999 losing tickets. More likely than not, the fans in the stadium will be disappointed (and it's more likely still they will be disappointed in a smaller stadium).

REVIEW QUESTIONS

1. Scientific notation is a format in which a number is expressed as a number between 1 and 10 multiplied by a power of 10. It makes writing numbers containing large numbers of zeros more compact. It makes approximations easier since we just need to add the exponents to account for very large or small numbers.

3. An order of magnitude estimate specifies a broad range of values, usually within one or two powers of ten, such as "in the ten thousands" or "in the millions." Even if the estimates are off by a factor of 10, they can still be used to compare the size of objects, such as the comparative size of a large or small city.

5. A scale can be described verbally in words such as "one centimeter represents one kilometer," shown graphically using marked miniruler on a map or diagram, or described as a ratio. On a map, a scale of 1 cm 1 km would be shown using a miniruler with lengths of 1 cm labeled as representing 100 km. The ratio would be 1 to 100,000,000 or 1/100,000,000, since there are 100,000,000 cm in 1 km.

7 As shown in Examples 7 and 8, the Earth would be about 1.3 millimeters in diameter and 15 meters from the sun. The three stars of the Alpha Centauri system would be 4100 km away.

DOES IT MAKE SENSE?

9. Makes sense. The page you are reading right now has two columns, each of which has about 50 lines, and there are roughly 10 words per line, for a total of 1000 words per page. A book with $10^5 = 100{,}000$ words would have 100 pages. A paperback with pages about half the size of this solutions manual would be two-hundred pages long.

11. Makes sense. At ten feet per floor, an office building with 30 floors would be 300 feet tall.

13. Does not make sense. This would mean the restaurant serves $5{,}000{,}000/365 = 13{,}699$ meals each day.

BASIC SKILLS AND CONCEPTS

15. a. $6\times10^4 = 60{,}000 =$ sixty thousand

b. $3\times10^5 = 300{,}000 =$ three hundred thousand

c. $3.4\times10^5 = 340{,}000 =$ three hundred forty thousand

d. $2\times10^{-3} = 0.002 =$ two thousandths

e. $6.7\times10^{-2} = 0.067 =$ sixty-seven thousandths

f. $3\times10^{-6} = 0.000003 =$ three millionths

17. a. $468 = 4.68\times10^2$

b. $126{,}547 = 1.26547\times10^5$

c. $0.04 = 4\times10^{-2}$

d. $9736.23 = 9.73623\times10^3$

e. $12.56 = 1.256\times10^1$

f. $0.8642 = 8.642\times10^{-1}$

19. a. $\left(3\times10^3\right)\times\left(2\times10^2\right) = 6\times10^5$

b. $\left(4\times10^5\right)\times\left(2\times10^5\right) = 8\times10^{10}$

c. $\left(3\times10^3\right)+\left(2\times10^2\right) = 3000+200$
$= 3200 = 3.2\times10^3$

d. $\left(6\times10^{10}\right)\div\left(3\times10^5\right) = 2\times10^5$

21. a. 10^{24} is 10^6 (1 million) times larger than 10^{18}.

b. 10^{27} is 10^{10} (10 billion) times larger than 10^{17}.

c. 1 trillion is 10^6 (1 million) times larger than 1 million.

23. My hard drive has a capacity of 1.2×10^{12} bytes.

25. The diameter of a typical atom is about 5×10^{-10} meter.

27. a. 260,000 is about 2.5×10^5 and 200 is 2×10^2, so their product is about 5×10^7, or 50,000,000. The exact value is 5.2×10^7.

b. 5.1 million is about 5×10^6 and 1.9 thousand is about 2×10^3, so their product is about 1×10^{10}, or 10 billion. The exact value is 9.69 billion.

27. (continued)

c. 600,000 is 6×10^5 and 3100 is about 3×10^3, so their quotient is about 2×10^2, or 200. A more precise value is 1.935×10^2, or 193.5.

29. Answers will vary regarding the estimates made, but most people spend more money on gas each month than on coffee. If you buy, on average, one coffee per day, spending $2 per coffee, that's around $60 per month. On the other hand, if you fill up your gas tank once a week, spending around $50 to fill a tank (assuming a 12 gallon tank and $4.00/gallon for gas,) that's around $200 to $250 per month.

31. Yes, assuming a quarter weighs about 0.2 ounces, the weight of $200 in quarters would be approximately $\$200\times\frac{4\text{ quarters}}{\$1}\times\frac{0.2\text{ oz}}{1\text{ quarter}}\times\frac{1\text{ lb}}{16\text{ oz}}=10\text{ lb.}$

33. $18 billion; 100 million pizza eaters $\times\frac{1\text{ pizza/person}}{1\text{ month}}\times\frac{\$15}{1\text{ pizza}}\times\frac{12\text{ months}}{1\text{ year}}$

35. 200 gallons per year; The recommended amount is 2 liters per day. $\frac{2\text{ liters}}{1\text{ day}}\times\frac{365\text{ days}}{1\text{ year}}\times\frac{0.264\text{ gallon}}{1\text{ liter}}$

37. 800 gallons; $\frac{20{,}000\text{ mi}}{1\text{ year}}\times\frac{1\text{ gallon}}{25\text{ mi}}$ (20,000 miles>yr and 25 mi>gal)

39. Table 3.1 states that 4 million joules are required for one hour of running, and that an average candy bar supplies 1 million joules of energy. Thus four candy bars would be needed for each hour of running, which means you'd need to eat 24 candy bars to supply the energy for six hours of running.

41. One kilogram of uranium-235 releases 5.6×10^{13} joules, which is 35,000 times as much as the energy released by one kilogram of coal (1.6×10^9 joules): $\frac{5.6\times10^{13}}{1.6\times10^9}=35{,}000.$

43. $10^4\text{ homes}\times\frac{5\times10^7\text{ J}}{\text{home}}\times\frac{1\text{ liter }H_2O}{6.9\times10^{13}\text{ J}}=7\times10^{-3}\text{liters;}$ This is about 0.007 liters.

45. $\frac{1.0\times10^{20}\text{ J}}{\text{yr}}\times\frac{1\text{ yr}}{365\text{ days}}\times\frac{1\text{ kg uranium}}{5.6\times10^{13}\text{ J}}=4.9\times10^3\frac{\text{kg}}{\text{day}}$; Thus 4900 kg of uranium would be needed to supply the daily energy needs of the United States.

47. 1 cm represents 20 km. Since there are 2,000,000 cm in 20 km, the scale is 2,000,000 to 1.

49. 1 cm represents 500 km. Since there are 50,000,000 cm in 500 km, the scale is 50,000,000 to 1.

51. In order to scale down each of the actual diameters and distances listed in the table, begin by dividing by 10,000,000,000 (10 billion). Since this produces tiny answers when units of kilometers are used, it is best to convert to a more convenient unit. The table below shows diameters listed in millimeters (rounded to the nearest tenth of a millimeter), and distances listed in meters (rounded to the nearest meter).

51. (continued)

Planet	Model Diameter	Model distance from Sun
Mercury	0.5 mm	6 m
Venus	1.2 mm	11 m
Earth	1.3 mm	15 m
Mars	0.7 mm	23 m
Jupiter	14.3 mm	78 m
Saturn	12.0 mm	143 m
Uranus	5.2 mm	287 m
Neptune	4.8 mm	450 m

53. a. The scale on this timeline would be 4.5 billion years to 100 meters, and thus 500 million years is equivalent to $0.5/4.5 = 1/9$ of 100 meters, which amounts to 11.1 meters.

b. Since 10,000 years is 1/50,000 of one billion years, we only need to divide 11.1 meters by 50,000 to find the distance that correlates with 10,000 years. This gives 0.000222 m, or $0.000222\text{ m}\times\frac{1000\text{ mm}}{1\text{ m}} = 0.222\text{ mm}$.

FURTHER APPLICATIONS

55. $\frac{2.69\times10^6\text{ deathss}}{\text{yr}}\times\frac{1\text{ yr}}{365\text{ d}}\times\frac{1\text{ d}}{24\text{ hr}}\times\frac{1\text{ hr}}{60\text{ min}} = 5.1\frac{\text{deaths}}{\text{min}}$

57. $\frac{101.5\times10^6\text{ passengers}}{\text{yr}}\times\frac{1\text{ yr}}{365\text{ d}}\times\frac{1\text{ d}}{24\text{ hr}} \approx 11{,}590\frac{\text{passengers}}{\text{hr}}$

59. $\frac{\$20\times10^{12}}{321\times10^6\text{ people}} \approx \$62{,}300\text{/person}$

61. $\frac{\$482\times10^9}{\text{yr}}\times\frac{1\text{ car}}{\$30{,}000}\times\frac{1\text{ yr}}{365\text{ d}}\times\frac{1\text{ d}}{24\text{ hr}}\times\frac{1\text{ hr}}{60\text{ min}} \approx 31\text{ cars/min}$

63. a. $1\text{ cm}^3\times\left(\frac{10^4\ \mu\text{m}}{1\text{ cm}}\right)^3\times\frac{1\text{ cell}}{100\ \mu\text{m}^3} = 1\times10^{10}$ cells, or 10 billion cells in 1 cm^3.

b. One liter is 1000 milliliters, or 1000 cubic centimeters, so there are 1000 times as many cells in a liter as there are in a cubic centimeter, which comes to 1×10^{13} cells (see part (a)).

c. If one liter of water weighs 1 kilogram, then a person who weighs 70 kg has a volume of 70 liters. From part (b), we know each liter contains 1×10^{13} cells, and thus we need only multiply that result by 70 to find the number of cells in a human body. This results in 7×10^{14} cells.

65. If we assume an average of three people per household, there are 100 million households in the U.S. (because there are 300 million people). If 4.5 trillion gallons is enough to supply all households for five months, then $\frac{4.5\times10^{12}}{5} = 9\times10^{11}$ gallons is sufficient for one month. Thus each household uses $\frac{9\times10^{11}\text{ gal}}{1\times10^8\text{ households}} = 9000$ gallons per month. Answers will vary regarding whether this is reasonable, but considering water used for bathing, washing, drinking, cooking, cleaning, and upkeep of the lawn/garden, this is within the ballpark.

67. a. The volume of a sphere is $V = \frac{4}{3}\pi r^3$, and thus the volume of the white dwarf will be $\frac{4}{3}\pi(6400\text{ km})^3$.

Density is mass per unit volume, so the density of the white dwarf will be

$$\frac{2\times10^{30}\text{ kg}}{\frac{4}{3}\pi(6400\text{ km})^3}\times\left(\frac{1\text{ km}}{10^5\text{ cm}}\right)^3 = 1821\frac{\text{kg}}{\text{cm}^3}.$$

b. Since a teaspoon is about 4 cubic centimeters, and each cubic centimeter has mass of 1821 kg (part (a)), a teaspoon will have a mass of 7300 kg, which is about the mass of a tank.

c. Following the calculations in part (a), the density is $\frac{2.8\times10^{30}\text{ kg}}{\frac{4}{3}\pi(10\text{ km})^3}\times\left(\frac{1\text{ km}}{10^5\text{ cm}}\right)^3 = 7\times10^{11}\frac{\text{kg}}{\text{cm}^3}$, which means one cubic centimeter of this material is more than ten times the mass of Mount Everest.

69. a. $\frac{\$12.3\times10^{12}\text{/year}}{3.21\times10^8\text{ people}} = \$38{,}300\text{/person/year}$

b. $\frac{\$38{,}300\text{/person}}{1\text{ year}}\times\frac{1\text{ year}}{365\text{ day}} = \$105\text{/person/day}$

c. $\frac{\$8.3\times10^{12}}{\$12.3\times10^{12}\text{/year}} \approx 0.67 = 67\%$

d. $\frac{\$2.1\times10^{12}}{\$12.3\times10^{12}\text{/year}} = 0.17 = 17\%$

e. Total spending: $\frac{\$12.3-\$6.8}{\$6.8} = 0.81 = 81\%$; Health care: $\frac{\$2100-\$918}{\$918} = 1.29 = 129\%$

71. Careful measurements show that a penny is about 1.5 mm thick, a nickel 1.8 mm, a dime 1.3 mm, and a quarter 1.7 mm thick. The easiest way to make these measurements is to measure a stack of ten coins, and divide the answer by ten. Although a quarter is almost the thickest coin (among these), its value is 2.5 times as much as the thinnest coin, the dime, and yet it's thickness is not anywhere near 2.5 times as thick as a dime (it's about 1.3 times as thick). So while you'll get more dimes in a stack that is as tall as you are, the value of that stack won't be as much as the value of the stack of quarters – take the quarters.

73. Answers will vary considerably, depending on the amount of light pollution in the area where one might make an estimate.

TECHNOLOGY EXERCISES

81. a. $1\text{ yr}\times186{,}000\frac{\text{mi}}{\text{sec}}\times\frac{60\text{ sec}}{1\text{ min}}\times\frac{60\text{ min}}{1\text{ hr}}\times\frac{24\text{ hr}}{1\text{ day}}\times\frac{365\text{ days}}{1\text{ yr}} = 5.87\times10^{12}\text{ miles}$

b. $\frac{52\cdot51\cdot50\cdot49\cdot48}{5\cdot4\cdot3\cdot2\cdot1} = 2{,}598{,}960\text{ hands}$

c. $\frac{37{,}000\times10^6\text{ metric tons}}{7.5\times10^9\text{ persons}} = 4.9\text{ metric tons per person}$

d. $$\frac{6.0\times10^{24}\text{ kg}\times\frac{1000\text{ g}}{1\text{ kg}}}{1.1\times10^{12}\text{ km}^3\times\left(\frac{100{,}000\text{ cm}}{1\text{ km}}\right)^3} = 5.5\frac{\text{g}}{\text{cm}^3}$$

e. $14\times10^9\text{ years}\times\frac{365\text{ days}}{1\text{ yr}}\times\frac{24\text{ hr}}{1\text{ day}}\times\frac{60\text{ min}}{1\text{ hr}}\times\frac{60\text{ sec}}{1\text{ min}} = 4.4\times10^{17}\text{ sec}$

UNIT 3C: DEALING WITH UNCERTAINTY

QUICK QUIZ

1. **b.** The best economists can do when making projections into the future is to base their estimates on current trends, but when these trends change, it often proves the predictions wrong.
2. **a.** There are two significant digits in 5.0×10^{-1}, while the other answers have only one.
3. **b.** 1.020 has four significant digits, 1.02 has three, and 0.000020 has only two.
4. **a.** Random errors can either be too high or too low, and averaging three readings is likely to reduce random errors because the high readings typically cancel the low readings.
5. **b.** If you place the thermometers in sunlight, this is a problem with your system of measurement, and the readings will all be too high, which is a systematic error.
6. **c.** Since all the scores are affected the same way (50 points too low), this is an example of a systematic error.
7. **b.** The absolute error in all cases was –50 points, but unless all students had the same score (very unlikely), the relative errors will be different.
8. **a.** Because the scale is able to report your weight to the nearest 1/10 of a pound, it is fairly precise (in comparison to standard bathroom scales), but its accuracy is lacking as it is off by more than 30 pounds (in 146 pounds).
9. **a.** Reporting the debt to the nearest penny is very precise, though the debt is so large that we really don't know it to that level of precision, and thus it is not likely to be very accurate.
10. **a.** Multiplying the gas mileage by the tank capacity will produce the number of miles the car can go on a full tank (290 miles), so the question boils down to this: "How many significant digits should be used when reporting the answer?" Since there are two significant digits in 29 mpg, and three in 10.0 gallons, we should use two significant digits in the answer (use the rounding rule for multiplication), and thus the answer is 290 miles.

REVIEW QUESTIONS

1. Significant digits in a number represent actual measurements. See the Information Box on page 156 for determining whether zeros are significant.
3. Absolute error describes how far a measured (or claimed) value lies from the true value. Relative error compares the size of the absolute error to the true value and is often expressed as a percentage. If you make seemingly large 500 pound error measuring the weight of a 200,000 pound ship, the relative error is only $\frac{500\text{ lb}}{200{,}000\text{ lb}} = 0.25\%$. A human hair is approximately 0.05 millimeters in diameter. An error of only 0.01 millimeter measuring the width of a hair would have a relative error of $\frac{0.01\text{ mm}}{0.05\text{ mm}} = 20\%$.
5. Giving measurements with more precision that justified can give them more certainty than they deserve.

DOES IT MAKE SENSE?

7. Does not make sense. Predicting the federal deficit for the next year to the nearest 100 million dollars is nearly impossible.
9. Does not make sense. Unless you have access to an exceptionally precise scale, you could never measure your height to that level of precision.
11. Makes sense. Suppose you measure your weight with a scale that is precise to the nearest 1/10 of a pound, but you fail to take your shoes and clothes off before stepping on the scale. The reading will not be very accurate (off by a pound or more), and increasing the level of precision to, say, 1/100 of a pound won't help matters.

BASIC SKILLS AND CONCEPTS

13. a. 6
b. 98
c. 0
d. 357
e. 12,784
f. 3
g. 7387
h. −16
i. −14

15. There are four significant digits, and it is precise to the nearest 1 (whole number).

17. There is one significant digit, and it is precise to the nearest hundred.

19. There are four significant digits, and it is precise to the nearest thousandth of a mile.

21. There are two significant digits, and it is precise to the nearest thousand seconds.

23. There are seven significant digits, and it is precise to the nearest ten-thousandth of a pound.

25. There are five significant digits, and it is precise to the nearest tenth of a km/s.

27. $23 \times 12.4 = 285$

29. $988 \div 10.3 = 96$

31. $\left(1.82 \times 10^3\right) \times \left(6.5 \times 10^{-2}\right) = 1.18 \times 10^2$

33. Random errors may occur when birds are miscounted, and systematic errors may happen the same bird is are counted more than once (if it leaves and then returns to the region) or a bird that enters the region is not seen.

35. Random errors could occur when taxpayers make honest mistakes or when the income amounts are recorded incorrectly. Systematic errors could occur when dishonest taxpayers report income amounts that are lower than their true income amounts.

37. Random errors could occur with an inaccurate radar gun or with honest mistakes made when the officer records the speeds. Systematic errors could occur with a radar gun that is incorrectly calibrated so it consistently reads too high or too low.

39. Problem (1) is a random error because the mistakes may be too high or two low. Problem (2) is a systematic error, because the tax payer likely underreported his income.

41. This is a systematic error that results in all altitude readings being around 500 feet higher than they should be.

43. The absolute error is $1.76 \text{ m} - 1.73 \text{ m} = 0.03$ meters. The relative error is $\frac{1.76 \text{ m} - 1.73 \text{ m}}{1.73 \text{ m}} = 0.017 = 1.7\%$.

45. The absolute error is $26 \text{ mi/hr} - 24 \text{ mi/hr} = 2 \text{ mi/hr}$. The relative error is $\frac{26 \text{ mi/hr} - 24 \text{ mi/hr}}{24 \text{ mi/hr}} = 0.083 = 8.3\%$.

47. The absolute error is $30 \text{ ft} - 33.5 \text{ ft} = -3.5$ feet. The relative error is $\frac{30 \text{ ft} - 33.5 \text{ ft}}{33.5 \text{ ft}} = -0.104 = -10.4\%$.

49. The absolute error is $97.9°\text{F} - 98.4°\text{F} = -0.5°\text{F}$. The relative error is $\frac{97.9°\text{F} - 98.4°\text{F}}{98.4°\text{F}} = -0.005 = 0.5\%$.

51. The measurement obtained by the tape measure is more accurate because the value is closer to your true height than the value obtained at the doctor's office (1/8 inch versus 0.4 inch). However, the laser at the doctor's office is more precise as it measures to the nearest 0.05 (or 1/20) inch, while the tape measure measures to the nearest 1/8 inch.

53. The digital scale at the gym is both more precise and more accurate. It is more precise because it measures to the nearest 0.01 kg, while the health clinic's scale measures to the nearest 0.5 kg, and it is more accurate because its reported weight is nearer to your true weight than the scale at the health clinic.

55. $136 \text{ lb} + 0.6 \text{ lb} = 136.6 \text{ lb}$; This should be rounded to the nearest 1 lb because that's the level of precision of 136 lb. The final answer is 137 lb.

57. 163 mi ÷ 2.3 hr ≈ 70.8696 mi/hr; Since 163 has three significant digits while 2.3 has only two, we must use only two in the final answer: 71 mi/hr.

59. The least precise measurement is 8 lb 4 oz, which is precise to the nearest ounce, so the answer should be rounded to the nearest ounce. $8 \text{ lb } 4 \text{ oz} + 14.6 \text{ oz} = 8 \text{ lb} \times \frac{16 \text{ oz}}{1 \text{ lb}} + 4 \text{ oz} + 14.6 \text{ oz} = 146.6 \text{ oz}$, which is 147 ounces when rounded to the correct level of precision.

61. The per capita cost is $\frac{\$2,800,000}{120,400 \text{ people}} \approx \$23.26/\text{person}$. However, since the first measurement of $2,800,000 has only two significant digits, whereas 120,400 has four significant digits, the answer must have two significant digits, so we round to $23 per person.

FURTHER APPLICATIONS

63. Random or systematic errors could be present; (for example, a visitor could be counted twice or not counted). Nevertheless, measurement is believable with the given precision.

65. Random or systematic errors could be present; (for example, a person could be counted twice or not counted). The measurement is not believable with the given precision.

67. Random or systematic errors could be present; also Asia is not well-defined. The measurement is not believable with the given precision.

69. Random or systematic errors could be present; (for example, are lost and stolen books included in the count?). The precision is a bit too high for the claim to be believable.

71. a. Since each cut can be off by 0.25 inches, the maximum error in either direction will be $20 \times 0.25 \text{ in} = 5 \text{ in}$. The lengths of the boards can be between $48 \text{ in} - 5 \text{ in} = 43 \text{ in}$ and $48 \text{ in} + 5 \text{ in} = 53 \text{ in}$.

b. Since each cut can be off by 0.5% or 0.005, the lengths of the boards can be between $48 \times (0.995)^{20} \approx 42.42$ inches and $48 \times (1.005)^{20} \approx 53.03$ inches.

UNIT 3D: INDEX NUMBERS: THE CPI AND BEYOND

QUICK QUIZ

1. **b.** The 2005 index of 127.9 is $\frac{192.5}{100} = 1.925$ times the 1985 index of 100, so the price of gas in 2005 is 1.925 times as much as the price in 1985.

2. **c.** If you know the price of gas in 1985, you can divide today's price by the 1985 price to get the index number for the current year.

3. **b.** Like most indexes, the CPI is designed to allow one to compare prices from one year to another.

4. **c.** To compare prices in one year to prices in another, one can just divide the CPI from one year by the other – this tells you the factor by which prices in one year differ from prices in another.

5. **c.** The CPI in 2008 was 215.3, and it was 152.4 in 1995. Thus prices in 2008 are 41% more than prices in 1995 (because $215.3/152.4 = 1.41$ If the CPI were recalibrated so that 1995 = 100, the new CPI in 2005 would be 141, which is lower than 215.3.

6. **b.** As long as the computers being compared are of comparable power, since the prices have declined, the index would decrease.

7. **a.** If the cost of college has increased faster than the CPI, but one's salary has only kept pace with the CPI, then the cost of college as a percentage of salary has increased, making it more difficult to afford.

8. **a.** If your salary increases faster than the CPI, then it has more purchasing power than in previous years.

9. **c.** Measured in 2006 dollars, gas was less than $1.50 per gallon in 1998–1999, and this is lower than any other time shown on the chart.

10. **c.** In order to reflect the fact that housing prices have tripled from 1985 to 2005, the index would also triple, from 100 to 300.

REVIEW QUESTIONS

1. Index numbers provide a simple way to compare measurements made at different times or in different places. The value at one particular time (or place) is chosen as the reference value. The index number for any other time (or place) is the ratio of the other value to the reference value.

3. Prices of most commodities tend to rise over time, so prices cannot be compared fairly unless the increase expected from inflation is taken into account. To adjust prices, the price in year Y is found by multiplying the price in year X by $\frac{\text{CPI}_{\text{year Y}}}{\text{CPI}_{\text{year X}}}$. $\frac{\text{CPI}_{\text{year Y}}}{\text{CPI}_{\text{year X}}}$ also indicates what \$1.00 in the year X would be worth in the year Y.

DOES IT MAKE SENSE?

5. Does not make sense. One must consider the effect of inflation on the price of gas by comparing prices using a price index of, say, 2015 dollars in order to make a valid comparison.

7. Makes sense. A penny in Franklin's day would be worth something on the order of a dollar today.

9. Does not make sense. The reference year that is chosen for an index number does not affect the trends of inflation, so if milk is more expensive in 1995 dollars, it should also be more expensive in 1975 dollars.

BASIC SKILLS AND CONCEPTS

11. To find the index number, take the current price and divide it by the 1985 price: $\$2.55/\$1.20 = 2.125 = 212.5\%$. The price index number is 212.5 (the percentage sign is dropped).

13. The price index number in 2015 is 210.0, which means prices in 2015 were 210.0% of the prices in 1985. Thus \$12 spent on gas in 1985 is equivalent to $\$12 \times 210.0\% = \$12 \times 2.100 = \$25.20$ in 2015 dollars.

15. To find the fraction of a tank you could purchase, divide the 1975 price by the 2015 price: $\frac{57.0¢}{252.0¢} = 0.23 = 23\%$, so you could buy 0.23 of the same tank with \$10 in 2015.

17. Divide the 2016 CPI by the 1986 CPI to find the factor by which prices in 2016 were larger than prices in 1986: $\frac{\text{CPI}_{2016}}{\text{CPI}_{1986}} = \frac{240.0}{109.6} = 2.190$. This means prices in 2016 were about 2.190 times larger than prices in 1986, so \$25,000 in 1986 is equivalent to $\$25{,}000 \times \frac{240.0}{109.6} = \$54{,}745$ in 2016.

19. Inflation is the relative change in the CPI, which is $\frac{\text{CPI}_{2005} - \text{CPI}_{1995}}{\text{CPI}_{1995}} = \frac{195.3 - 152.4}{152.4} \approx 0.281 = 28.1\%$. Note that this is inflation over the span of ten years (that is, it is not an annual rate of inflation).

21. The price in 2010 dollars is $\$0.25 \times \frac{\text{CPI}_{2010}}{\text{CPI}_{1977}} = \$0.25 \times \frac{218.1}{60.6} = \$0.90$.

23. The price in 1980 dollars is $\$10.00 \times \frac{\text{CPI}_{1980}}{\text{CPI}_{2016}} = \$10.00 \times \frac{82.4}{240.0} = \$3.43$.

25. The purchasing power of a 1977 dollar in terms of 2010 dollars is $\$1 \times \frac{\text{CPI}_{2010}}{\text{CPI}_{1977}} = \$1 \times \frac{218.1}{60.6} = \$3.60$.

FURTHER APPLICATIONS

27. In Miami, the price would be $\$400{,}000 \times \frac{161}{122} = \$528{,}000$.

29. In San Francisco, the price would be $\$300{,}000 \times \frac{172}{121} = \$426{,}000$.

31. Health care spending increased by $\frac{\$3.2\times10^{12} - \$85\times10^{9}}{\$85\times10^{9}} = 36.65 = 3665\%$, , whereas the overall rate of inflation was $\frac{CPI_{2015} - CPI_{1977}}{CPI_{1977}} = \frac{237.0-60.6}{60.6} = 2.91 = 291\%$.

33. The relative change in the cost of private colleges was $\frac{\$33{,}480 - \$8396}{\$8396} = 2.988 = 298.8\%$. The rate of inflation was $\frac{CPI_{2016} - CPI_{1990}}{CPI_{1990}} = \frac{240.0-130.7}{130.7} = 0.836 = 83.6\%$.

35. As shown in the first line of the table, \$0.25 in 1938 dollars is worth \$2.78 in 1996 dollars, so \$1 in 1938 dollars will be worth $4\times\$2.78 = \$11.12$ in 1996 dollars.

37. In 1996, actual dollars are 1996 dollars.

39. The 1979 minimum wage, measured in 1979 dollars, was \$2.90. To convert this to 1996 dollars, use the CPI from 1996 and 1979: $\$2.90\times\frac{CPI_{1996}}{CPI_{1979}} = \$2.90\times\frac{156.9}{72.6} = \$6.27$; This agrees with the entry in the table.

41. The purchasing power of the minimum wage was highest in 1968, when its value in 1996 dollars was highest (\$7.21). Since all entries in the last column are listed in 1996 dollars, we can make valid comparisons between values in the column.

43. a. An index is usually the ratio of two quantities with the same units (such as prices), which means it has no units.

b. The major league average is \$219.53, so the index values for each team are:

Team	Index
Boston	$\frac{\$360.66}{\$219.53}\times100 = 164.3$
New York (Yankees)	$\frac{\$337.20}{\$219.53}\times100 = 153.6$
Chicago (Cubs)	$\frac{\$312.32}{\$219.53}\times100 = 142.3$
Colorado	$\frac{\$193.96}{\$219.53}\times100 = 88.4$
San Diego	$\frac{\$182.82}{\$219.53}\times100 = 83.3$
Arizona	$\frac{\$132.10}{\$219.53}\times100 = 60.2$
Major League Average	$\frac{\$219.53}{\$219.53}\times100 = 100.0$

45. Answers will vary depending on when the index was last recalculated.

TECHNOLOGY EXERCISES

51. \$100 in 1980 had the same buying power as \$260 in 2009.

53. \$25 in 1930 had the same buying power as \$321 in 2009.

55. *A* will be greater than *B*.

UNIT 3E: HOW NUMBERS DECEIVE: POLYGRAPHS, MAMMOGRAMS, AND MORE

QUICK QUIZ

1. **a.** 8 refers to the number of patients with severe acne that were cured by the new treatment.
2. **a.** As stated in the text surrounding the table, the old treatment had a better overall treatment rate, even though the new treatment performed better in each category.
3. **b.** Even though Derek had a higher GPA in each of the two years, we can't be sure his overall GPA was higher than Terry's – Simpson's paradox may be at play (see historical note on page 178).
4. **a.** A *false negative* means the test detected no cancer (that's the *negative* part) even though the person has it (the *false* part).
5. **c.** A *false positive* means the test detected steroids (that's the *positive* part), but the person never used steroids (that's the *false* part).
6. **c.** The last row in the column labeled *Tumor is benign* shows the total of those who did not have malignant tumors.
7. **b.** Those who had false negatives (15) and false positives (1485) make up the group who had incorrect test results.
8. **c.** Imagine a test group of 1000 women, where 900 are actually pregnant, and 100 are not. Of those who are pregnant, 1% (that is, 9 women) will test negative. Of those who aren't pregnant, 99% (that is, 99 women) will test negative. Thus 108 women test negative, and 99 of them actually aren't pregnant. Since 99/108 is not 99% (it's closer to 92%), statement c is not true (it may be true in other scenarios, but in general, it is not).
9. **c.** Taxpayers in the top 1% would have paid 31% of tax revenue without the tax cuts and paid 37% of tax revenue with the tax cuts, which most likely means they paid more in income taxes.
10. **b.** The figure of $1900 is a taxpayers' savings, not their annual tax payment.

REVIEW QUESTIONS

1. False positives occur when a test indicates a true result when the actual outcome should be false. False negatives occur when a test indicates a negative result when the actual outcome should be true. True positives occur when a test indicates a positive result when the actual outcome is also true. True negatives occur when a test indicates a negative result when the actual outcome is also false.

3. If a polygraph, or drug test, is 90% accurate, even when every person is either telling the truth or is not using drugs, out of 1000 people tested, $10\% \times 1000 = 100$ of them could be accused of lying or using drugs.

DOES IT MAKE SENSE?

5. Makes sense. Both categories could show an improvement with the new drug even though overall, the old drug may have done a better job.

7. Does not make sense. There are scenarios (see discussion in text concerning mammograms and drug tests, for example) where a positive test may not correlate well with the probability that a bag contains banned materials, even though the test is 98% accurate.

9. Does not make sense. Both sides can make a valid argument for their position (see discussion in text about tax cuts).

BASIC SKILLS AND CONCEPTS

11. a. As shown in the tables, Josh had a higher batting average in the first half of the season.

    b. As shown in the tables, Josh had a higher batting average in the second half of the season.

    c. Jude had the higher overall batting average $(80/200 = .400 \text{ versus } 85/220 = .386)$.

    d. One person (Josh) performed better in two of two categories (the first and second halves of the season), and yet Jude outperformed Josh over the course of the entire season.

13. a. New Jersey had higher scores in both categories (283 versus 281, and 252 versus 250), but Nebraska had the higher overall average (277 versus 272).

b. It's possible for one state (Nebraska) to score lower in both categories, and yet have a higher overall percentage. This is due to the different racial makeup of the two state's populations. The white students are scoring better than the nonwhite students in both states, and Nebraska has a larger percentage of white students than New Jersey, so they influence the overall scores more heavily than the white students in New Jersey.

c. The verification is an exercise in a weighted average. 87% of the population had an average of 281, and 13% of the population had an average of 250, so the overall average is $0.87(281)+0.13(250)=277$.

d. As in part (c), we only have to compute a weighted average: $0.66(283)+0.34(252)=272$.

e. See part (b).

15. a. The death rate for whites was $8400/4,675,000=0.0018=0.18\%$. For nonwhites, it was $500/92,000=0.0054=0.54\%$. Overall, the death rate was $8900/4,767,000=0.0019=0.19\%$.

b. The death rate for whites was $130/81,000=0.0016=0.16\%$. For nonwhites, it was $160/47,000=0.0034=0.34\%$. Overall, the death rate was $290/128,000=0.0023=0.23\%$.

c. The death rates for TB in each category (white and nonwhite) were higher in New York City than in Richmond, and yet the overall death rate due to TB was higher in Richmond. The paradox arises because Richmond had a different racial makeup than New York City: it had a much higher percentage of nonwhite residents than New York, and as the death rate for TB was much higher among nonwhites than whites in both cities, the higher percentage of nonwhites in Richmond has a more pronounced influence on the death rate.

17. a. There are 10,000 total participants with tumors in the study (lower right hand entry), and 1% of them (100) have malignant tumors. Since the test is 90% accurate, 90 of those with malignant tumors will be true positives, while the other 10 will be false negatives. 99% of the women (9900) will have benign tumors. 90% of these (8910) will be true negatives, while 10% of these (990) will be false positives. The only numbers in the table not yet verified are the first two entries in the Total column; these are simply the sum of the rows in which they lie.

b. Of the 1080 who have positive mammograms, only 90 are true positives, so only $90/1080=8.3\%$ of them actually have cancer.

c. 100 of the women have cancer (malignant tumors), and 90 of those will have a true positive test result, which is 90%. Thus if you really have cancer, the test will detect it 90% of the time.

d. Of the 8920 who have a negative mammogram, only 10 of them are false negatives, which translates to a probability of $10/8920=0.11\%$.

19. a. Beginning with the last row, the sample size is 2000, and 2% of them (40) used drugs, while the remaining 1960 did not. The first column shows the 40 drug users, and the test detects 90% of them (36), but misses 4 of them. The second column shows the 1960 that used drugs, and the test says 90% of them (1764) did not use drugs (a negative test), but it says 10% of them (196) did use drugs. The entries in the last column are simply the sum of the rows in which they lie.

b. 232 of the athletes were accused of using drugs. 36 of the 232 used drugs and while 196 of the 232 (84.5%) were falsely accused of using drugs.

c. 1768 of the athletes were cleared of using drugs. 4 of the 1768 (0.2%) who were cleared actually used drugs.

21. One can argue that funding for the program is being cut because there was a 1% rise in the budget, and yet the CPI is 3%, which means the purchasing power of the new budget has decreased. On the other hand, the funding has certainly increased from \$1 billion to \$1.01 billion.

FURTHER APPLICATIONS

23. a. Spelman won 10/29 = 34.5% of its home games and 12/16 = 75% of its away games, while Morehouse won 9/28 = 32.1% of its home games and 56/76 = 73.7% of its away games. Since Spelman is better than Morehouse in each category, it's the better team.

b. Morehouse won 65/104 = 62.5% of all the games they played, while Spelman won 22/45 = 48.9% of all the games they played. Based on these figures, Morehouse is the better team.

23. (continued)

c. Disregarding the strength of the opponents each college played, it's universally accepted that the team with a higher percentage of wins is the better team, and thus the claim in part (b) makes more sense. Additionally, it's more difficult to win away games than home games, and while both teams have an impressive away-game win percentage, Morehouse compiled its overall record of 62.5% with many more away games.

25. a. Note that 10% of the 5000 at risk people is 500, and this is the sum of the first row in the first table. For the 20,000 people in the general population, note that 0.3% of 20,000 is 60, and this is the sum of the first row in the second table. 95% of the 500 infected people in the at risk group should test positive, and 5% should test negative – take 95% and 5% of 500 to verify the first row entries in the first table. 95% of the 4500 at risk people who aren't infected should test negative, while 5% should test positive – take 95% and 5% of 4500 to verify the second row in the first table.

A similar process will show that the numbers in the second table are also correct.

b. 475/500 = 95% of those with HIV test positive. 475/700 = 67.9% of those who test positive have HIV. The percentages are different because the number of people in each group (those who have HIV and those who test positive) is different.

c. As shown in part (b), if you test positive, and you are in the at risk group, you have a 67.9% chance of carrying the disease. This is much greater than the overall percentage of those who have the disease (10%), which means if you test positive, you should be concerned.

d. 57/60 = 95% of those with HIV test positive. 57/1054 = 5.4% of those who test positive have HIV. The two percentages are different because the number of people in each group (those who have HIV and those who test positive) is different.

e. As shown in part (d), if you test positive, and you are in the general population, you have a 5.4% chance of carrying the disease. This is larger than the overall percentage of those who have the disease (0.3%), but it's still small enough that it's best to wait for more tests before becoming alarmed.

27. a. Excelsior Airlines has a higher percentage of on-time flights in each of the five cities shown in the table.

b. First, compute the number of overall on-time arrivals for Excelsior by multiplying the percents in each row by the number of arrivals for each city, and adding these together:

0.889(559) + 0.948(233) + 0.914(232) + 0.831(605) + 0.858(2146) = 3274 on-time flights. Divide this by the total number of flights (3775) to find the overall on-time percentage: 3274/3775 = 86.7%.

Next, do the same thing for Paradise:

0.856(811) + 0.921(5255) + 0.855(448) + 0.713(449) + 0.767(262) = 6438 on-time flights, which means an overall on-time percentage of 6438/7225 = 89.1%.

c. Excelsior had a higher percentage of on-time flights in each of the five cities, yet Paradise had the higher overall percentage of on-time flights.

UNIT 4A: TAKING CONTROL OF YOUR FINANCES

QUICK QUIZ

1. **a.** Evaluating your budget allows you to look critically at your cash flow, which affects personal spending.
2. **a.** Your cash flow is determined by the amount of money you earn (income), and the amount you spend.
3. **b.** If your cash flow is negative, you are spending more than you take in.
4. **c.** You should prorate all once-per-year expenses and include them in your monthly budget.
5. **b.** Housing typically costs about one-third of your income.
6. **c.** As a percentage of income, health care expenses are not too alarming, though they grow rapidly as one ages.
7. **a.** You can't save without money left over at the end of the month, which corresponds to a positive cash flow.
8. **a.** The bill for the collision is greater than her deductible, so Sandy pays \$500 out-of-pocket.
9. **a.** Without any medical bills, you only pay the premium of \$4800.
10. **c.** After the deductible and co-payment are applied, the remaining balance is $\$2700-\$1500-\$200=\$1000$. The insurance company pays 80% of the \$1000, which is \$800.

REVIEW QUESTIONS

1. Understanding your finances helps you avoid suffering from financial stress, no matter your level of income. It also helps you keep better track of your spending, making going into debt less likely. You can also better plan to have the necessary money for college, purchasing a house, and retirement.

3. Making a budget means keeping track of how much money you have coming in and going out and then deciding what adjustments you need to make. Cash flow is the amount of money remaining after subtracting total expenses from total income. The four-step process is:
Step 1. Determine your average monthly income.
Step 2. Determine your average monthly expenses.
Step 3. Determine your net monthly cash flow by subtracting your total expenses from your total income.
Step 4. Make adjustments as needed.

5. The premium is the amount you pay to purchase the policy. A deductible is the amount you are personally responsible for before the insurance company will pay anything. A co-payment usually applies to health insurance and is the amount you pay each time you use a particular service, such as on office visit, that is covered by the insurance policy. When evaluating the benefits of an insurance policy, you should consider the policy's maximum benefits, exceptions that may lead to a lack of coverage, and your potential cost if you don't purchase coverage.

DOES IT MAKE SENSE?

7. Does not make sense. All the smaller expenses do add up, and they have a significant influence on your budget.

9. Makes sense. When prorated as a per-month expense, an \$1800 vacation costs \$150 per month.

11. Does not make sense. Divide \$15,000 by 365 (days per year) to find that an annual expense of \$15,000 comes to \$41 per day. Pizza and a soda do not cost that much.

BASIC SKILLS AND CONCEPTS

13. Maria spends \$25 each week on coffee, and using 52 weeks per year, this comes to \$1300 per year. She spends \$150 each month on food, which is \$1800 per year. The total yearly cost is \$3100. The amount she spends on coffee is $\$1300/\$1800=72\%$ of the amount she spends on food.

15. Suzanne spends \$85 per month for her cell phone usage, which is \$1020 per year. The total yearly cost is \$1220. The amount she spends on her cell phone is $\$1020/\$200=510\%$ of the amount she spends on health insurance.

17. Sheryl spends \$12 each week on cigarettes, which comes to \$624 per year. She spends \$40 each month on dry cleaning, or \$480 per year. The total yearly cost is \$1104. The amount she spends on cigarettes is $\$624/\$480 = 130\%$ of the amount she spends on dry cleaning.

19. Vern spends \$27 per week on beer, which amounts to \$1404 per year. This is $\$1404/\$800 = 176\%$ of the amount he spends on textbooks. The total yearly cost is \$2204.

21. Since $18\%/12 = 1.5\%$, you spend 1.5% of \$750 on interest each month, which is $0.015 \times \$750 = \$11.25$. Over the span of a year, you'll spend \$135 on interest.

23. Sam's balance is $\$2800 - \$400 = \$2400$. He spends 3% of \$2400, or $0.03 \times \$2400 = \$72$ each month on interest. This comes to \$864 per year.

25. Riley spends \$7100 per semester, or \$14,200 per year, which is $\$14{,}200/12 = \$1183.33$ per month.

27. Talib spends $\$750 \times 2 = \$1500$ per year on automobile insurance, $\$150 \times 12 = \$1800$ per year for health insurance, and \$500 per year on life insurance. The total yearly cost is \$3800, which means he spends $\$3800/12 = \$316.67$ per month.

29. Finn spends $\$250 + \$300 + \$150 + \$450 = \$1150$ each year on these contributions, or $\$1150/12 = \$95.83$ per month.

31. Total income: $\$650 \times 12 + \$400 \times 12 + \$6000 = \$18{,}600$ per year.
Total Expenses: $\$500 \times 12 + \$60 \times 48 + \$3600 \times 2 + \$120 \times 48 = \$21{,}840$ per year.
Annual cash flow: $\$18{,}600 - \$21{,}840 = -\$3240$
Monthly cash flow: $-\$3240/12 = -\$270$

33. Total income: $\$1900 \times 12 = \$22{,}800$ per year.
Total Expenses: $\$800 \times 12 + \$90 \times 48 + \$125 \times 12 + \$150 \times 12 + \$420 \times 2 + \$25 \times 48 + \$400 \times 12 = \$24{,}060$ per year.
Annual cash flow: $\$22{,}800 - \$24{,}060 = -\$1260$
Monthly cash flow: $-\$1260/12 = -\$105$

35. The woman spends $\$900/\$3600 = 25\%$ of her income on rent (i.e. housing), which is below average.

37. The man spends $\$200/\$4200 = 5\%$ on health care, which is slightly below average.

39. The couple spends $\$800/\$5200 = 15\%$ on health care, which is slightly below average.

41. The cost of the old car is computed below.

Gas: $\frac{250 \text{ mi}}{\text{wk}} \times \frac{1 \text{ gal}}{21 \text{ mi}} \times \frac{\$3.50}{1 \text{ gal}} \times \frac{52 \text{ wk}}{\text{yr}} \times 5 \text{ yr}$
$= \$10{,}833$

Insurance: $\frac{\$400}{\text{yr}} \times 5 \text{ yr} = \$2000$

Repairs: $\$1500 \times 5 = \$7500$

Total: $\$10{,}833 + \$2000 + \$7500 = \$20{,}333$

The cost of the new car is computed below.

Gas: $\frac{250 \text{ mi}}{\text{wk}} \times \frac{1 \text{ gal}}{45 \text{ mi}} \times \frac{\$3.50}{1 \text{ gal}} \times \frac{52 \text{ wk}}{\text{yr}} \times 5 \text{ yr}$
$= \$5056$

Insurance: $\frac{\$800}{\text{yr}} \times 5 \text{ yr} = \$4000$

Purchase Price: \$16,000

Total: $\$5056 + \$4000 + \$16{,}000 = \$25{,}056$

Using the old car is less expensive, but not by a large factor. After five years, if you use the old car, you'll have a junky car, whereas if you bought the new car, you would have a nice car after five years.

43. If you buy the car for \$22,000, and sell it three years later for \$10,000, you will have spent \$12,000. If you lease the car for three years (36 months) you will spend $\$1000 + \$250 \times 36 = \$10{,}000$. It is cheaper to lease in this case.

45. At the in-state college, you will spend $\$4000 + \$700 \times 12 = \$12{,}400$ each year. At the out-of-state college, you will spend $\$6500 + \$450 \times 12 = \$11{,}900$. It will cost less out-of-state.

47. He will earn $(\$14,988-\$9012)\times 12\times 40=\$2,868,480$ more over 40 years.

49. A man earns approximately $\dfrac{\$14,988-\$11,580}{\$11,580}=29\%$ more than a woman in a year. Over 40 years, a man will earn $(\$14,988-\$11,580)\times 12\times 40=\$1,635,840$ more.

51. If you take the course, you will spend \$1500, and use 150 hours of time that could be used to earn money in a job. If you work those 150 hours, you will earn \$1500. Therefore, the net cost will be $\$1500+\$1500=\$3000$.

53. a. You pay the premiums and the deductible, for out-of-pocket expenses of $\$650\times 2+\$1000=\$2300$.

b. You pay for everything, for a total cost of $\$1120+\$1660=\$2780$.

55. Plan A: You pay only the premium and deductible, for a total of $\$5400+\$400=\$5800$.
Plan B: You pay the premium, deductible, and co-payments (\$230), for a total of $\$1500+\$2000+\$230=\$3730$. You also pay 20% of the remaining balance, which is $20\%\times(\$3500+\$300+\$700-\$2000-\$230)=\$454$. Your out-of-pocket expenses are $\$3730+\$454=\$4184$.

FURTHER APPLICATIONS

57. Note that it costs \$10 less per month to operate the new dryer, and thus in 62 months, you would have saved \$620, which is long enough to pay for the new dryer.

59. a. If you didn't have the policy, you would pay all of the costs of the claims, which is $\$450+\$925=\$1375$. If you did have the policy, you would pay two years of premiums $(2\times\$550=\$1100)$, all of the \$450 claim, and \$500 of the \$925 claim, for a grand total of \$2050.

b. Without the insurance policy, your cost would be $\$450+\$1200=\$1650$. With the policy, you would pay premiums for two years, and \$200 on each claim for a total of $2\times\$650+2\times\$200=\$1700$.

c. Without the policy, your cost would be $\$200+\$1500=\$1700$. With the policy, you would pay premiums for two years, \$200 for the first claim, and \$1000 for the second claim, for a total of $2\times\$300+\$200+\$1000=\$1800$.

61. Under Plan A, you'll pay \$1000 for the down payment, \$400 for each of 24 months (two years), and \$10,000 for the residual, which totals to $\$1000+\$400\times 24+\$10,000=\$20,600$. Under Plan B, the cost will be $\$500+\$250\times 36+\$9500=\$19,000$. Under Plan C, the cost will be $\$175\times 48+\$8000=\$16,400$, and this is the least expensive plan.

63. – 65. Answers will vary.

UNIT 4B: THE POWER OF COMPOUNDING

QUICK QUIZ

1. **b.** Compound interest always yields a greater balance than simple interest when the APR is the same.

2. **a.** If you begin with a principle P, and add 6% to its value after one year, the result is $P+0.06P=1.06P$, which means the principle increases by a factor of 1.06 each year.

3. **c.** The compound interest formula states $A=P\times(1+\text{APR})^Y$. With an APR of 5.5%, the balance after five years would be $A=P\times(1+0.055)^5$, which means the account increases in value by a factor of 1.055^5.

4. **a.** The APR of 4% is divided evenly into four parts so that each quarter (three months), the account increases in value by $4\%/4=1\%$.

5. **c.** Compounding interest more often always results in a greater annual percentage yield.

6. **b.** The APR is the same as the APY with annual compounding, but it's smaller in all other cases.

7. **a.** After 20 years, the account earning 4% APR will have grown by a factor of $1.04^{20} = 2.19$, whereas the account earning 2% APR will have grown by a factor of $1.02^{20} = 1.49$. Thus the 4% APR account will have earned about one and a half times as much interest.

8. **a.** The continuously compounded interest formula reads $A = P \times e^{(\text{APR} \times Y)}$, and thus after two years, the balance is $A = \$250 \times e^{(0.06 \times 2)} = \$250 \times e^{0.12}$.

9. **a.** The compound interest formula assumes a constant APR for as many years as you have your money invested in an account.

10. **c.** Bank accounts that earn compound interest grow by larger amounts as time goes by, and this is a hallmark of exponential growth.

REVIEW QUESTIONS

1. For simple interest, interest is only calculated using the principle. For compound interest, interest is calculated on both the principle and any interest already earned. Since the balance on which compound interest is calculated increases over time, you are earning interest on a larger amount than just the principle.

3. APR/n is the interest rate at each payment if interest is paid n times per year. The APR is evenly divided between all the payments.

5. $A = P\left(1 + \frac{\text{APR}}{n}\right)^{(nY)}$, where A = accumulated balance after Y years, P = starting principal, APR = annual percentage rate (as a decimal), n = number of compounding periods per year, and Y = number of years.

7. When interest is compounded infinitely many times per year, it is called continuous compounding. It represents the best possible compounding for a particular APR, so it will have a higher APY than any other form of compound interest. As the number of compoundings increases, the value of $\left(1 + \frac{\text{APR}}{n}\right)^{(nY)}$ approaches $e^{(\text{APR} \times Y)}$, so the formula for continuous compounding only requires the starting principal, P, the APR, as a decimal, and Y, the number of years, to calculate the final balance.

DOES IT MAKE SENSE?

9. Does not make sense. You earn more interest under compound interest when the APR is the same.

11. Does not make sense. A bank with an APR of 5.9% compounded daily is a better deal than a bank with 6% APR compounded annually (all other things being equal) because the APY in the first case is greater.

13. Makes sense. One's bank account grows based on the APY, and an APR of 5% could certainly result in an APY of 5.1% if the number of compounding periods per year was just right.

BASIC SKILLS AND CONCEPTS

15. $3^2 = 3 \times 3 = 9$

17. $2^5 = 2 \times 2 \times 2 \times 2 \times 2 = 32$

19. $25^{1/2} = \sqrt{25} = 5$

21. $64^{-1/3} = \frac{1}{64^{1/3}} = \frac{1}{\sqrt[3]{64}} = \frac{1}{4}$

23. $3^4 \div 3^2 = 3^{4-2} = 3^2 = 9$

25. $25^{1/2} \div 25^{-1/2} = 25^{1/2-(-1/2)} = 25^1 = 25$

27. $x = 20$ (Add 4 to both sides.)

29. $z = 16$ (Add 10 to both sides.)

31. $y = 4$ (Divide both sides by 4.)

33. $z = 4$ (Add 1, and divide both sides by 5.)

35. $3x - 4 = 2x + 6 \Rightarrow 3x = 2x + 10 \Rightarrow x = 10$

37. $3a + 4 = 6 + 4a \Rightarrow 3a - 2 = 4a \Rightarrow a = -2$

39. $6q-20=60+4q \Rightarrow 6q=80+4q \Rightarrow 2q=80 \Rightarrow q=40$

41. $t/4+5=25 \Rightarrow t/4=20 \Rightarrow t=80$

43. $x^2=25 \Rightarrow (x^2)^{1/2}=25^{1/2} \Rightarrow x=\sqrt{25}=5$; Another solution is $x=-5$.

45. $(x-4)^2=36 \Rightarrow \left((x-4)^2\right)^{1/2}=36^{1/2} \Rightarrow x-4=\sqrt{36} \Rightarrow x=4+6=10$; Another solution is $x=2$.

47. $(t/3)^2=16 \Rightarrow \left((t/3)^2\right)^{1/2}=16^{1/2} \Rightarrow t/3=\sqrt{16} \Rightarrow t=3\times4=12$; Another solution is $t=-12$.

49. $u^9=512 \Rightarrow (u^9)^{1/9}=512^{1/9} \Rightarrow u=\sqrt[9]{512}=2$

51. You'll earn 5% of \$800, or $0.05\times\$800=\$40$, each year. After five years, you will have earned $5\times\$40=\$200$ in interest, so that your balance will be \$1000.

53. Each year, you'll earn 3.5% of \$3200, or $0.035\times\$3200=\$112$. After five years, you will have earned $5\times\$112=\$560$ in interest, and your balance will be \$3760.

55.

	Suzanne		Derek	
Year	Interest	Balance	Interest	Balance
0	–	\$3000	–	\$3000
1	\$75	\$3075	\$75	\$3075
2	\$75	\$3150	\$77	\$3152
3	\$75	\$3225	\$79	\$3231
4	\$75	\$3300	\$81	\$3312
5	\$75	\$3375	\$83	\$395

Suzanne's balance increases by \$375, or 12.5% of its original value. Derek's increases by \$395, or 13.2% of its original value. (Note on values shown in the table: the balance in one year added to next year's interest does not necessarily produce next year's balance due to rounding errors. Each value shown is rounded correctly to the nearest cent).

57. $A=\$5000(1+0.04)^{10}=\$7401.22$

59. $A=\$15,000(1+0.032)^{25}=\$32,967.32$

61. $A=\$10,000(1+0.037)^{12}=\$15,464.83$

63. $A=\$5000\left(1+\frac{0.02}{4}\right)^{4\times10}=\$6103.97$

65. $A=\$25,000\left(1+\frac{0.03}{365}\right)^{365\times5}=\$29,045.68$

67. $A=\$4000\left(1+\frac{0.06}{12}\right)^{12\times20}=\$13,240.82$

69. $A=\$25,000\left(1+\frac{0.037}{4}\right)^{4\times30}=\$75,472.89$

71. The APY is the relative increase in the balance of a bank account over the span of one year. The easiest way to compute it when the APR is known is to find the one-year balance, compute the relative increase, and express the answer as a percent. When the principal is not given, any amount can be used – here, we will use \$100.

1-year balance: $A=\$100\left(1+\frac{0.041}{365}\right)^{365\times1}=\$104.18$. The relative increase is $\$4.18/\$100=4.18\%$, so the APY is 4.18%.

73. See Exercise 71 for details. The one-year balance is $A=\$100\left(1+\frac{0.0123}{12}\right)^{12\cdot1}=\$101.24$. The relative increase (APY) is $\$1.24/\$100=1.24\%$.

75. The one, five, and twenty-year balances are:

One year: $A = \$5000e^{0.045\times1} = \$5230.14$

Five years: $A = \$5000e^{0.045\times5} = \$6261.61$

Twenty years: $A = \$5000e^{0.045\times20} = \$12,298.02$ $20,137.53

To compute the APY, find the relative increase in the balance of the account after one year.

$\text{APY} = \$230.14/\$5000 = 4.60\%$

77. The one, five, and twenty-year balances are:

One year: $A = \$7000e^{0.045\times1} = \$7322.20$

Five years: $A = \$7000e^{0.045\times5} = \$8766.26$

Twenty years: $A = \$7000e^{0.045\times20} = \$17,217.22$

To compute the APY, find the relative increase in the balance of the account after one year.

$\text{APY} = \$322.20/\$7000 = 4.60\%.$

79. The one, five, and twenty-year balances are:

One year: $A = \$3000e^{0.06\times1} = \$3185.51$

Five years: $A = \$3000e^{0.06\times5} = \$4049.58$

Twenty years: $A = \$3000e^{0.06\times20} = \$9960.35$

To compute the APY, find the relative increase in the balance of the account after one year.

$\text{APY} = \$185.51/\$3000 = 6.18\%$

81. Solve $\$25,000 = P(1+0.06)^8$ for P to obtain $P = \dfrac{\$25,000}{(1+0.06)^8} = \$15,685.31.$

83. Solve $\$25,000 = P\left(1+\dfrac{0.06}{12}\right)^{12\times8}$ for P to obtain $P = \dfrac{\$25,000}{\left(1+\dfrac{0.06}{12}\right)^{12\times8}} = \$15,488.10.$

85. Solve $\$120,000 = P(1+0.055)^{15}$ for P to obtain $P = \dfrac{\$120,000}{(1+0.055)^{15}} = \$53,751.97.$

87. Solve $\$120,000 = P\left(1+\dfrac{0.028}{4}\right)^{4\times15}$ for P to obtain $P = \dfrac{\$120,000}{\left(1+\dfrac{0.028}{4}\right)^{4\times15}} = \$78,961.07.$

FURTHER APPLICATIONS

89. After 10 years, Chang has $705.30; after 30 years, he has $1403.40. After 10 years, Kio has $722.52; after 30 years, she has $1508.74. Kio has $17.22, or 2.4% more than Chang after 10 years. She has $105.34, or 7.5% more than Chang after 30 years.

91. To compute the APY, find the one-year balance in the account, using any principal desired ($100 is used here), and then compute the relative increase in the value of the account over one year.

One-year balance, compounded quarterly:

$$A = \$100\left(1+\frac{0.053}{4}\right)^{4\times 1} = \$105.41$$

APY = $5.41/$100 = 5.41%

One-year balance, compounded monthly:

$$A = \$100\left(1+\frac{0.053}{12}\right)^{12\times 1} = \$105.43$$

APY = $5.43/$100 = 5.43%

One-year balance, compounded daily:

$$A = \$100\left(1+\frac{0.053}{365}\right)^{365\times 1} = \$105.44$$

APY = $5.44/$100 = 5.44%

As the number of compounding periods per year increases, so does the APY, though the rate at which it increases slows down.

93.

	Account 1		Account 2	
Year	Interest	Balance	Interest	Balance
0	–	$1000		$1000
1	$55	$1055	$57	$1057
2	$58	$1113	$59	$1116
3	$61	$1174	$63	$1179
4	$65	$1239	$67	$1246
5	$68	$1307	$70	$1317
6	$72	$1379	$74	$1391
7	$76	$1455	$79	$1470
8	$80	$1535	$83	$1553
9	$84	$1619	$87	$1640
10	$89	$1708	$93	$1733

Account 1 has increased in value by $708, or 70.8%. Account 2 has increased by $733, or 73.3%. (Note on values shown in the table: the balance in one year added to next year's interest does not necessarily produce next year's balance due to rounding errors. Each value shown is rounded correctly to the nearest dollar).

95. a. After 5 years, Rosa has $\$3000(1.04)^5 = \$3649.96$; after 20 years, she has $\$3000(1.04)^{20} = \$6573.37$. Julian has $\$2500(1.05)^5 = \$3190.70$ after 5 years, and he has $\$2500(1.05)^{20} = \$6633.24$ after 20 years.

b. For Rosa, $\frac{\$3649.96-\$3000.00}{\$3649.96} = 18\%$ of the balance after 5 years is interest and $\frac{\$6573.37-\$3000.00}{\$6573.37} = 54\%$ of the balance after 20 years is interest. For Julian, $\frac{\$3190.70-\$2500.00}{\$3190.70} = 22\%$ of the balance after 5 years is interest and $\frac{\$6633.24-\$2500.00}{\$6633.24} = 62\%$ of the balance after 20 years is interest.

c. The longer the investment time, the higher the overall returns.

97. For Plan A, solve $\$120{,}000 = P(1.05)^{30}$ for P to find $P = \frac{\$120{,}000}{(1.05)^{30}} = \$27{,}765.29$.

For Plan B, solve $\$120{,}000 = Pe^{0.048\times 30}$ for P to find $P = \frac{\$120{,}000}{e^{0.048\times 30}} = \$28{,}431.33$.

99. Solve $3P = P(1.06)^t$ by dividing both sides by P to obtain $3 = (1.06)^t$ and using trial and error (substituting values for t) to obtain $t = 18.9$ years.
To obtain an exact solution for this problem, one must use logarithms (studied in a chapter 8). To solve $3P = P(1.06)^t$ for t, begin by dividing both sides by P (it will cancel from both sides), and then take the logarithm of both sides, which results in $\log_{10} 3 = \log_{10} 1.06^t$. A property of logarithms says the exponent t can be moved in front of the logarithm, as such: $\log_{10} 3 = t \times \log_{10} 1.06$. Now divide both sides by $\log_{10} 1.06$ to arrive at the answer: $t = \dfrac{\log_{10} 3}{\log_{10} 1.06} \approx 18.9$ years. Thus it will take about 18.9 years for your investment to triple in value.

101. Solve $\$100{,}000 = \$1000(1.07)^t$ by trial and error (substituting values for t) to obtain $t = 68.1$ years. Using the method described in Exercise 99 results in the exact value of $t = \dfrac{\log_{10} 100}{\log_{10} 1.07} = 68.1$ years.

103. a. The account balance would be $A = \$1{,}500{,}000{,}000\left(1+\dfrac{0.06}{12}\right)^{12\times 1} = \$1{,}592{,}516{,}718$ at the end of the year, and the interest would be $\$92{,}516{,}718$.

b. Yes, the account balance would be $A = \$1{,}500{,}000{,}000\left(1+\dfrac{0.05}{12}\right)^{12\times 1} = \$1{,}576{,}742{,}847$ at the end of the year, and the interest of $\$76{,}742{,}847$ is sufficient for the total donation of $75,000,000.

c. About $\dfrac{\$1{,}500{,}000{,}000}{5\text{ yr}} \times \dfrac{1\text{ yr}}{52\text{ weeks}} = 5{,}769{,}230$, or approximately $5.8 million per week.

TECHNOLOGY EXERCISES

109. a. FV(0.10,5,0,100) = $161.05

b. FV(0.02,535,0,224) = $8,940,049.24

111. a. FV(0.03/12,60,0,5000) = $5808.08

b. FV(0.045/12,30*12,0,800) = $3078.16

c. FV(0.0375/365,365*50,0,1000) = $6520.19

113. a. $e^{3.2} = 24.5325$

b. $e^{0.065} = 1.0672$

c. $100 \times e^{0.04} = 104.0811$; So the APY is $\$4.081/\$100 = 4.0811\%$.

UNIT 4C: SAVINGS PLANS AND INVESTMENTS

QUICK QUIZ

1. a. As the number of compounding periods n increases, so does the APY, and a higher APY means a higher accumulated balance.
2. c. As long as the savings plan in which you are investing money has a positive rate of return, the balance will increase as time increases.
3. c. The total return on a five-year investment is simply the percent change of the balance over five years.
4. b. The annual return is the APY that gives the same overall growth in five years as the total return.

5. **a.** If you deposit \$200 per month for five years (60 months), you have deposited a total of \$12,000 into the savings plan. Since the balance was \$22,200, you earned $\$22,200 - \$12,000 = \$10,200$ in interest.

6. **a.** If you can find a low risk, high return investment, that's the best of both worlds, and adding high liquidity (where it's easy to access your money) is like icing on the cake. Good luck.

7. **b.** Market capitalization is the share price of a company times the number of outstanding shares. Company B has the greatest market capitalization of $\$9$ per share $\times 10,000,000$ shares $= \$90,000,000$, which is greater than the market capitalization of \$10,000,000 for companies A and C.

8. **a.** The P/E ratio is the share price divided by earnings per share over the past year.

9. **b.** If the bond is selling at 103 points, it is selling at 103% of its face value, which is $1.03 \times \$5000 = \$5150$.

10. **c.** The one-year return is simply a measure of how well the mutual fund has done over the last year, and it could be higher or lower than the three-year return (which measures how well the fund has done over the last three years).

REVIEW QUESTIONS

1. Instead of one, large initial deposit, a savings plan makes deposits of a smaller amount on a regular basis. The savings plan formula (See the Information Box on page 226.) allows you to calculate the final balance (future value) of a series of payments, given the payment amount, the number of deposits per year, and the length of the payment plan.

3. Examples will vary. Consider an investment that grows from an original principal P to a later accumulated balance A. The total return is the percentage change in the investment value, so total return $= \dfrac{(A-P)}{P} \times 100\%$.

   The annual return is the annual percentage yield (APY) that would give the same overall growth over Y years, so annual return $= \left(\dfrac{A}{P}\right)^{(1/Y)} - 1$.

5. Liquidity represents how difficult is it to take out your money. An investment from which you can withdraw money easily, such as an ordinary bank account, is said to be liquid. The liquidity of an investment like real estate is much lower because real estate can be difficult to sell.

   Risk represents the risk to your investment principal. The safest investments are federally insured bank accounts and U.S. Treasury bills, where there is virtually no risk of losing the principal you've invested. Stocks and bonds are much riskier because they can drop in value, in which case you may lose part or all of your principal.
   Return represents how much you can expect to earn on your investment. A higher return means you earn more money. In general, low-risk investments offer relatively low returns, while high-risk investments offer the prospects of higher returns and the possibility of losing your principal.

7. The face value (or par value) of the bond is the price you must pay the issuer to buy it at the time it is issued. The coupon rate of the bond is the simple interest rate that the issuer promises to pay. The maturity date of the bond is the date on which the issuer promises to repay the face value of the bond. The current yield of a bond is the amount of interest it pays each year divided by the bond's current price (not its face value). A bond selling at a discount from its face value has a current yield that is higher than its coupon rate. The reverse is also true: A bond selling at a premium over its face value has a current yield that is lower than its coupon rate.

DOES IT MAKE SENSE?

9. Does not make sense. With an APR of 4%, the savings plan formula gives a balance of

$$A=\frac{\$25\left[\left(1+\frac{0.04}{12}\right)^{12\cdot 30}-1\right]}{\left(\frac{0.04}{12}\right)}=\$17{,}351$$ after 30 years, which probably wouldn't cover expenses for even one year of retirement (and certainly interest earned on that balance would not be enough for retirement). You should plan to save considerably more than \$25 per month for a retirement plan.

11. Does not make sense. Stocks can be a risky investment (you might lose all of your money), and most financial advisors will tell you to diversify your investments.

13. Does not make sense. Stocks are considered a risky investment (especially in the short term). In general, an investment with a greater return is even more risky, and thus the claim that there is no risk at all is dubious, at best.

BASIC SKILLS AND CONCEPTS

15. $$A=\frac{\$75\left[\left(1+\frac{0.03}{12}\right)^{12}-1\right]}{\left(\frac{0.03}{12}\right)}=\$912.48$$

17. $$A=\frac{\$200\left[\left(1+\frac{0.04}{12}\right)^{12\times 3}-1\right]}{\left(\frac{0.04}{12}\right)}=\$7636.31$$

19. You save for 40 years, so the value of the IRA is $A=\frac{\$150\left[\left(1+\frac{0.05}{12}\right)^{12\times 40}-1\right]}{\left(\frac{0.05}{12}\right)}=\$228{,}903.02$. Since you deposit \$150 each month for 40 years, your total deposits are $\frac{\$150}{\text{mo}}\times\frac{12\text{ mo}}{\text{yr}}\times 40\text{ yr}=\$72{,}000$. The value of the account is just over three times the amount you deposited.

21. $A=\frac{\$300\left[\left(1+\frac{0.035}{12}\right)^{12\times 18}-1\right]}{\left(\frac{0.035}{12}\right)}=\$90{,}091.51$; You have deposited $\frac{\$300}{\text{mo}}\times\frac{12\text{ mo}}{\text{yr}}\times 18\text{ yr}=\$64{,}800$, which is a little less than three-fourths of the value of the account.

23. Solve $\$170{,}000=\frac{PMT\left[\left(1+\frac{0.05}{12}\right)^{12\times 15}-1\right]}{\left(\frac{0.05}{12}\right)}$ for PMT. $PMT=\frac{\$170{,}000\left(\frac{0.05}{12}\right)}{\left[\left(1+\frac{0.05}{12}\right)^{12\times 15}-1\right]}=\$636.02$; You should deposit \$636.02 each month.

25. Solve $\$30{,}000=\frac{PMT\left[\left(1+\frac{0.055}{12}\right)^{12\times 3}-1\right]}{\left(\frac{0.055}{12}\right)}$ for PMT. $PMT=\frac{\$30{,}000\left(\frac{0.055}{12}\right)}{\left[\left(1+\frac{0.055}{12}\right)^{12\times 3}-1\right]}=\$768.38$; You should deposit \$768.38 each month.

27. First, find out how large your retirement account needs to be in order to produce $100,000 in interest (at 6% APR) each year. You want 6% of the total balance to be $100,000, which means the total balance should be $100,000/0.06 = $1,666,667.

Now solve $\$1,666,667 = \dfrac{PMT\left[\left(1+\frac{0.06}{12}\right)^{12\times30}-1\right]}{\left(\frac{0.06}{12}\right)}$ for *PMT*. $PMT = \dfrac{\$1,666,667\left(\frac{0.06}{12}\right)}{\left[\left(1+\frac{0.06}{12}\right)^{12\times30}-1\right]} = \$1659.18$. You should deposit $1659.18 each month.

29. The total return is the relative change in the investment, and since you invested $6000 (100 shares at $60 per share), the total return is $\dfrac{\$9400-\$6000}{\$6000} = 0.567 = 56.7\%$. The annual return is the APY that would give the same overall growth in five years, and the formula for computing it is $\left(\dfrac{A}{P}\right)^{(1/Y)} - 1$. Thus the annual return is $\left(\dfrac{\$9400}{\$6000}\right)^{(1/5)} - 1 = 0.094 = 9.4\%$

31. The total return is $\dfrac{\$11,300-\$6500}{\$6500} = 73.8\%$. The annual return is $\left(\dfrac{\$11,300}{\$6500}\right)^{(1/20)} - 1 = 2.8\%$.

33. The total return is $\dfrac{\$2000-\$3500}{\$3500} = -42.9\%$. The annual return is $\left(\dfrac{\$2000}{\$3500}\right)^{(1/3)} - 1 = -17.0\%$.

35. The total return is $\dfrac{\$12,600-\$7500}{\$7500} = 68.0\%$. The annual return is $\left(\dfrac{\$12,600}{\$7500}\right)^{(1/10)} - 1 = 5.3\%$.

37. $300 invested in stocks would be worth $\$300(1+0.064)^{116} = \$400,267$. For bonds and cash, the investments would be worth $\$300(1+0.020)^{116} = \$2984$ and $\$300(1+0.008)^{116} = \$756$, respectively.

39. a. The symbol for Intel stock is INTC.
    b. Intel stock closed at $35.48 per share yesterday.
    c. The total value of shares traded today is $\$35.48/\text{share} \times 10,368,000 = \$367,856,640$, or about $368 million.
    d. $\dfrac{10.4 \text{ million}}{4709 \text{ million}} = 0.22\%$ of all Intel shares have been traded so far today.
    e. You should expect a dividend of $100 \times \$35.48 \times 0.0307 = \$108.92$.
    f. The earnings per share is $\dfrac{\$35.48}{15.4} = \$2.304$, or $2.30/share.
    g. Intel earned a total profit of $\$2.304/\text{share} \times 4709$ million shares $= \$10,849$ million, so Intel earned a total profit of about $10.85 billion.

41. a. The earnings per share for COSTCO is $\$171.60/20.84 = \$8.23/\text{share}$.
    b. The stock is slightly overpriced.

43. a. The earnings per share for IBM is $\$151.98/12.49 = \$12.17/\text{share}$.
    b. The stock is priced just about right.

45. Answers will vary.

47. The current yield on a bond is its annual interest payment divided by its current price. A \$1000 bond with a coupon rate of 2% pays $\$1000 \times 0.02 = \$20$ per year in interest, so this bond has a current yield of $\$20/\$950 = 2.11\%$.

49. The annual interest payment on this bond is $\$1000 \times 0.055 = \$55$, so its current yield is $\$55/\$1100 = 5\%$.

51. The price of this bond is $105\% \times \$1000 = \$1050$, and it has a current yield of 3.9%, so the annual interest earned is $\$1050 \times 0.039 = \$40.95$.

53. The price of this bond is $114.3\% \times \$1000 = \$1143$, and its current yield is 6.2%, which means the annual interest will be $\$1143 \times 0.062 = \$70.87$.

55. a. You would buy $\$5000/\$10.99 = 454.96$ shares.

b. Your investment would be worth $\$5000 \times (1+0.0125)^5 = \$5320.41$.

c. Your investment would be worth $\$5000 \times (1+0.0245)^{10} = \$6369.27$.

FURTHER APPLICATIONS

57. After 10 years, the balance in Yolanda's account is $A = \dfrac{\$200\left[\left(1+\dfrac{0.05}{12}\right)^{12\times10}-1\right]}{\left(\dfrac{0.05}{12}\right)} = \$31{,}056.46$. She deposits \$200 per month, so her total deposits are $\dfrac{\$200}{\text{mo}} \times \dfrac{12\text{ mo}}{\text{yr}} \times 10\text{ yr} = \$24{,}000$. Zach's account is worth $A = \dfrac{\$2400\left[\left(1+\dfrac{0.05}{1}\right)^{1\times10}-1\right]}{\left(\dfrac{0.05}{1}\right)} = \$30{,}186.94$. He deposits \$2400 per year, so his total deposits are $\dfrac{\$2400}{\text{yr}} \times 10\text{ yr} = \$24{,}000$. Yolanda comes out ahead even though both deposited the same amount of money because the interest in her account is compounded monthly, while Zach's is compounded yearly, which means Yolanda enjoys a higher APY.

59. After 10 years, the balance in Juan's account is $A = \dfrac{\$400\left[\left(1+\dfrac{0.06}{12}\right)^{12\times10}-1\right]}{\left(\dfrac{0.06}{12}\right)} = \$65{,}551.74$. His total deposits are $\dfrac{\$400}{\text{mo}} \times \dfrac{12\text{ mo}}{\text{yr}} \times 10\text{ yr} = \$48{,}000$. The balance in Maria's account is $A = \dfrac{\$5000\left[(1+0.065)^{10}-1\right]}{(0.065)} = \$67{,}472.11$. Her total deposits are $\dfrac{\$5000}{\text{yr}} \times 10\text{ yr} = \$50{,}000$. Maria comes out ahead because she has a higher APR (and APY, it turns out), and she deposits more money over the course of the 10 years.

61. Your balance will be $A = \dfrac{\$50\left[\left(1+\dfrac{0.07}{12}\right)^{12\times15}-1\right]}{\left(\dfrac{0.07}{12}\right)} = \$15{,}848.11$, so you won't reach your goal.

63. Your balance will be $A = \frac{\$100\left[\left(1+\frac{0.06}{12}\right)^{12\times15}-1\right]}{\left(\frac{0.06}{12}\right)} = \$29{,}081.87$, so you won't reach your goal.

65. The total return is $\frac{\$8.25-\$6.05}{\$6.05} = 36.4\%$ per share. Incidentally, this is also the annual return (because you bought the stock a year ago), and it doesn't matter how many shares you bought.

67. The total return was $\frac{\$22{,}000{,}000-\$5000}{\$5000} = 439{,}900\%$. The annual return was $\left(\frac{\$22{,}000{,}000}{\$5000}\right)^{(1/50)}-1$ $= 18.3\%$, which is much higher than the average annual return for large-company stocks.

69. a. At age 35, the balance in Mitch's account is $A = \frac{\$1000\left[(1+0.05)^{10}-1\right]}{0.05} = \$12{,}577.89$. At this point, it is no longer appropriate to use the savings plan formula as Mitch is not making further deposits into the account. Instead, use this balance as the principal in the compound interest formula, and compute the balance 40 years later. $A = \$12{,}578(1+0.05)^{40} = \$88{,}548.20$.

b. At age 75, the balance in Bill's account is $A = \frac{\$1000\left[(1+0.05)^{30}-1\right]}{0.05} = \$66{,}438.85$.

c. Mitch deposited \$1000 per year for ten years, which is \$10,000. Bill deposited \$1000 per year for 30 years, which is \$30,000, or three times as large as Mitch's total deposits.

d. Answer will vary.

TECHNOLOGY EXERCISES

77. a. FV(0.04/12,12*25,100,0) = \$51,412.95

b. FV(0.08/12,12*25,100,0) = \$95,102.64; The amount is less than double

c. FV(0.04/12,12*50,100,0) = \$190,935.64; The amount is more than double.

79. a. $2.8^{1/4} = 1.293569$

b. $120^{1/3} = 4.932424$

c. We need to solve the equation $1850 = 250\times(1+r)^{15}$. Using algebra, rewrite the equation as $\frac{1850}{250} = (1+r)^{15}$. Now raise both sides to the reciprocal power of 15, i.e. to the 1/15 power, and then subtract 1 to find $r = \left(\frac{1850}{250}\right)^{1/15} - 1 = 0.1427 = 14.27\%$.

UNIT 4D: LOAN PAYMENTS, CREDIT CARDS, AND MORTGAGES

QUICK QUIZ

1. **a.** Monthly payments go up when the loan principal is larger because you are borrowing more money.
2. **a.** The payment will be higher for a 15-year loan because there is less time to pay off the principal.
3. **b.** A higher APR means more money goes to paying off interest, and this corresponds with a higher payment.
4. **b.** Most of an early payment goes to interest, because the principal is large at the beginning stages of a loan, and thus the interest is also large.
5. **c.** Every year you'll pay \$12,000, and thus after ten years, you'll pay \$120,000.

6. **c.** Most credit card loans only require that you make a minimum payment each month; there is no specified time in which you must pay off a loan.

7. **c.** A two-point origination fee just means that you must pay 2% of the loan principal in advance, and 2% of $200,000 is $4000.

8. **b.** Add $500 to 1% (one point) of the loan principle of $120,000 to find the advanced payment required.

9. **c.** Refinancing a loan is not a good idea if you've nearly paid the loan off.

10. **a.** A shorter loan always has higher monthly payments (all other things being equal) because there is less time to pay off the principal, and you'll spend less on interest because the principal decreases more quickly.

REVIEW QUESTIONS

1. If you only pay the interest on a loan, the principle will never be reduced. Therefore, the loan will never be paid off.

3. As time progresses, the portion of a payment going to interest will decrease and the portion of a payment going towards the principle will increase.

5. Credit card loans differ from installment loans in that you are not required to pay off your balance in any set period of time. Instead, you are required to make only a minimum monthly payment that generally covers all the interest but very little principal. As a result, it takes a very long time to pay off your credit card loan if you make only the minimum payments.

DOES IT MAKE SENSE?

7. Makes sense. For a typical long-term loan, most of the payments go toward interest for much of the loan term. See Figure 4.12 in the text.

9. Does not make sense. Making only the minimum required payment on a credit card is asking for financial trouble as the interest rates on credit cards are typically exorbitant. You'll end up spending a lot of your money on interest.

11. Makes sense. In the worst-case scenario, the ARM loan will be at 6% in the last year you expect to live in the home, so it's certain you'll save money on the first two years, and you may even save during the last year.

BASIC SKILLS AND CONCEPTS

13. a. The starting principal is $120,000 with an annual interest rate of 6%. You'll make 12 payments each year (one per month) for 15 years, and each payment will be $1013.

 b. Since you make 12 payments per year for 15 years, you'll make $12\times15=180$ payments in total, which amounts to $180\times\$1013=\$182,340$ over the term of the loan.

 c. The loan principal is $120,000/$182,340 = 65.8% of the total payment. $\$182,340-\$120,000=\$62,340$, so $\$62,340/\$182,340=34.2\%$ of the total payment is interest.

15. a. The monthly payment is $PMT=\dfrac{\$100,000\left(\dfrac{0.06}{12}\right)}{1-\left(1+\dfrac{0.06}{12}\right)^{-12\times20}}=\$716.43$.

 b. The total payment is $\$716.43\times12\times20=\$171,943.20$.

 c. The percent spent on interest is $\dfrac{\$171,943.20-\$100,000}{\$171,943.20}=0.418=41.8\%$, while $\dfrac{\$100,000}{\$171,943.20}=0.582=58.2\%$ goes toward the principal.

17. a. The monthly payment is $PMT = \dfrac{\$400,000\left(\dfrac{0.035}{12}\right)}{1-\left(1+\dfrac{0.035}{12}\right)^{-12\times30}} = \$1796.18.$

b. The total payment is $\$1796.18\times12\times30 = \$646,624.80.$

c. The percent spent on interest is $\dfrac{\$646,624.80-\$400,000}{\$646,624.80} = 0.381 = 38.1\%,$ while

$\dfrac{\$400,000}{\$646,624.80} = 0.619 = 61.9\%$ goes toward the principal.

19. a. The monthly payment is $PMT = \dfrac{\$100,000\left(\dfrac{0.03}{12}\right)}{1-\left(1+\dfrac{0.03}{12}\right)^{-12\times15}} = \$690.58.$

b. The total payment is $\$\$690.58\times12\times15 = \$124,304.40.$

c. The percent spent on interest is $\dfrac{\$124,304.40-\$100,000}{\$124,304.40} = 0.196 = 19.6\%,$ while

$\dfrac{\$100,000}{\$124,304.40} = 0.804 = 80.4\%$ goes toward the principal.

21. a. The monthly payment is $PMT = \dfrac{\$10,000\left(\dfrac{0.08}{12}\right)}{1-\left(1+\dfrac{0.08}{12}\right)^{-12\times3}} = \$313.36.$

b. The total payment is $\$313.36\times12\times3 = \$11,2801.$

c. The percent spent on interest is $\dfrac{\$11,281-\$10,000}{\$11,281} = 0.114 = 11.4\%,$ while $\dfrac{\$10,000}{\$11,281} = 0.886 = 88.6\%$ goes toward the principal.

23. a. The monthly payment is $PMT = \dfrac{\$150,000\left(\dfrac{0.05}{12}\right)}{1-\left(1+\dfrac{0.05}{12}\right)^{-12\times15}} = \$1186.19.$

b. The total payment is $\$1186.19\times12\times15 = \$213,514.$

c. The percent spent on interest is $\dfrac{\$213,514-\$150,000}{\$213,514} = 0.297 = 29.7\%,$ while

$\dfrac{\$150,000}{\$213,514} = 0.703 = 70.3\%$ goes toward the principal.

25. The monthly payment is $PMT = \dfrac{\$150{,}000\left(\dfrac{0.04}{12}\right)}{1-\left(1+\dfrac{0.04}{12}\right)^{-12\times 30}} = \$716.12.$

To compute the interest owed at the end of one month, multiply the beginning principal of \$150,000 by the monthly interest rate of 4%/12. This gives $\$150{,}000(0.04/12) = \$500.00.$ Because the monthly payment is \$716.12, the amount that goes toward principal is $\$716.12 - \$500.00 = \$216.12.$ Subtract this from \$150,000 to get a new principal of $\$150{,}000 - \$216.12 = \$149{,}783.88.$

For the second month, repeat this process, but begin with a principal of \$149,783.88. The results are shown in the following table through the third month – note that rounding errors at each step propagate through the table, which means that our answers may differ slightly from those that a bank would compute.

End of month	Interest	Payment toward principal	New Principal
1	\$500.00	\$216.12	\$149,783.88
2	\$499.28	\$216.84	\$149,567.04
3	\$498.56	\$217.56	\$149,349.48

27. For the 3-year loan at 7% APR, the monthly payment is $PMT = \dfrac{\$15{,}000\left(\dfrac{0.07}{12}\right)}{1-\left(1+\dfrac{0.07}{12}\right)^{-12\times 3}} = \$463.16.$ The monthly payment for the 4-year loan is \$362.68, and for the 5-year loan, it is \$304.15 (the computations being similar to those used for the 3-year loan). Since you can afford only the payments on the 5-year loan, that loan best meets your needs.

29. a. The monthly payment is $PMT = \dfrac{\$5000\left(\dfrac{0.18}{12}\right)}{1-\left(1+\dfrac{0.18}{12}\right)^{-12\times 1}} = \$458.40.$

b. Your total payments are $\$458.40 \times 12 = \$5500.80.$

c. The interest is $\$500.80/\$5500.80 = 9.1\%$ of the total payments.

31. a. The monthly payment is $PMT = \dfrac{\$5000\left(\dfrac{0.21}{12}\right)}{1-\left(1+\dfrac{0.21}{12}\right)^{-12\times 3}} = \$188.38.$

b. Your total payments are $\$188.38 \times 36 = \$6781.68.$

c. The interest is $\$1781.68/\$6781.68 = 26.3\%$ of the total payments.

33. Complete the table as shown in the first month. To find the interest for the second month, compute $\$1093.00 \times 0.015 = \$16.40$. To find the new balance, compute $\$1093 + \$75 + \$16.40 - \$200 = \$984.40$. Continue in this fashion to fill out the rest of the table. Note that rounding errors at each step will propagate through the table: a credit card company computing these balances might show slightly different values.

Month	Payment	Expenses	Interest	New balance
0				$1200.00
1	$200	$75	$18.00	$1093.00
2	$200	$75	$16.40	$984.40
3	$200	$75	$14.77	$874.17
4	$200	$75	$13.11	$762.28
5	$200	$75	$11.43	$648.71
6	$200	$75	$9.73	$533.44
7	$200	$75	$8.00	$416.44
8	$200	$75	$6.25	$297.69
9	$200	$75	$4.47	$177.16
10	$200	$75	$2.66	$54.82

In the 11th month, a partial payment of $\$54.82 + \$54.82 \times 0.015 + \$75 = \$130.64$ will pay off the loan.

35. Complete the table as shown in the first month. To find the interest for the second month, compute $\$179.50 \times 0.015 = \$2.69$. To find next month's balance, compute $\$179.50 + \$150 + \$2.69 - \$150 = \$182.19$. Continue in this fashion to fill out the rest of the table. Note that rounding errors at each step will propagate through the table: a credit card company computing these balances might slow slightly different values.

Month	Payment	Expenses	Interest	New balance
0				$300.00
1	$300	$175	$4.50	$179.50
2	$150	$150	$2.69	$182.19
3	$400	$350	$2.73	$134.92
4	$500	$450	$2.02	$86.94
5	$0	$100	$1.30	$188.24
6	$100	$100	$2.82	$191.06
7	$200	$150	$2.87	$143.93
8	$100	$80	$2.16	$126.09

37. Option 1: The monthly payment is $PMT = \dfrac{\$400,000\left(\dfrac{0.08}{12}\right)}{1-\left(1+\dfrac{0.08}{12}\right)^{-12\times 30}} = \$2935.06$, and the total cost of the loan is $\$2935.06 \times 12 \times 30 = \$1,056,621.60$.

Option 2: The monthly payment is $PMT = \dfrac{\$400,000\left(\dfrac{0.075}{12}\right)}{1-\left(1+\dfrac{0.075}{12}\right)^{-12\times 15}} = \$3708.05$, and the total cost of the loan is $\$3708.05 \times 12 \times 15 = \$667,449.00$.

You'll pay considerably more interest over the term of the loan in Option 1, but you may be able to afford the payments under that option more easily.

39. Option 1, the monthly payment is $PMT = \dfrac{\$60{,}000\left(\dfrac{0.0715}{12}\right)}{1-\left(1+\dfrac{0.0715}{12}\right)^{-12\times 30}} = \$405.24,$ and the total cost of the loan is $\$405.24 \times 12 \times 30 = \$145{,}886.40.$

Option 2, the monthly payment is $PMT = \dfrac{\$60{,}000\left(\dfrac{0.0675}{12}\right)}{1-\left(1+\dfrac{0.0675}{12}\right)^{-12\times 15}} = \$530.95,$ and the total cost of the loan is $\$530.95 \times 12 \times 15 = \$95{,}571.00.$

You'll pay considerably more interest over the term of the loan in Option 1, but you may be able to afford the payments under that option more easily.

41. Under Choice 1, your monthly payment is $PMT = \dfrac{\$120{,}000\left(\dfrac{0.04}{12}\right)}{1-\left(1+\dfrac{0.04}{12}\right)^{-12\times 30}} = \$572.90,$ and the closing costs are \$1200.

For Choice 2, the payment is $PMT = \dfrac{\$120{,}000\left(\dfrac{0.035}{12}\right)}{1-\left(1+\dfrac{0.035}{12}\right)^{-12\times 30}} = \$538.85,$ and the closing costs are $\$1200 + \$120{,}000(0.02) = \$3600.$ You'll save $\$572.90 - \$538.85 = \$34.05$ each month with Choice 2, though it will take $\$3600/(\$34.05 \text{ per month}) = 106$ months (about 9 years) to recoup the higher closing costs, and thus Choice 2 is the better option only if you intend to keep the house for at least 9 years.

43. Under Choice 1, your monthly payment is $PMT = \dfrac{\$120{,}000\left(\dfrac{0.045}{12}\right)}{1-\left(1+\dfrac{0.045}{12}\right)^{-12\times 30}} = \$608.02,$ and the closing costs are $\$1200 + \$120{,}000(0.01) = \$2400.$ For Choice 2, the payment is $PMT = \dfrac{\$120{,}000\left(\dfrac{0.0425}{12}\right)}{1-\left(1+\dfrac{0.0425}{12}\right)^{-12\times 30}} = \$590.33,$ and the closing costs are $\$1200 + \$120{,}000(0.03) = \$4800.$ You'll save $\$608.02 - \$590.33 = \$17.69$ each month with Choice 2, though it will take $\$2400/(\$17.29 \text{ per month}) = 139$ months (12 years) to recoup the higher closing costs, and thus Choice 2 is the better option only if you intend to keep the house for at more than 11 years.

45. a. The monthly payment is $PMT = \dfrac{\$30{,}000\left(\dfrac{0.09}{12}\right)}{1-\left(1+\dfrac{0.09}{12}\right)^{-12\times 20}} = \$269.92.$

45. (continued)

b. The monthly payment is $PMT = \dfrac{\$30{,}000\left(\dfrac{0.09}{12}\right)}{1-\left(1+\dfrac{0.09}{12}\right)^{-12\times10}} = \$380.03.$

c. If you pay the loan off in 20 years, the total payments will be $\$269.92\times12\times20=\$64{,}780.80$. If you pay it off in 10 years, the total payments will be $\$380.03\times12\times10=\$45{,}603.60$.

47. Following example 9 in the text, the interest on the \$150,000 loan in the first year will be approximately $4.5\%\times\$150{,}000=\$6750$, which means the monthly payment will be about $\$6750/12=\$562.50$. With the 3% ARM, the interest will be approximately $3\%\times\$150{,}000=\$4500$, and thus the monthly payment will be about $\$4500/12=\$375$. You'll save $\$562.50-\$375=\$187.50$ each month with the ARM. In the third year, the rate on the ARM will be 2.0 percentage points higher than the fixed rate loan, and thus the yearly interest payments will differ by $\$150{,}000\times0.02=\$3000$, which amounts to \$250 per month.

FURTHER APPLICATIONS

49. Solve $\$500 = \dfrac{P\left(\dfrac{0.0375}{12}\right)}{1-\left(1+\dfrac{0.0375}{12}\right)^{-12\times30}}$ for P by computing the values in the numerator and denominator on the right, and isolating P. This results in $P=\$107{,}964$ (rounded to the nearest dollar), and it represents the largest loan you can afford with monthly payments of \$500. The loan will pay for 80% of the house you wish to buy (you must come up with a 20% down payment). Thus $80\%\times(\textit{house price})=\$107{,}964$, which means $\textit{house price}=\$107{,}964/0.80=\$134{,}956$. You can afford a house that will cost about \$134,956.

51. a. The monthly payment for the first loan is $PMT = \dfrac{\$10{,}000\left(\dfrac{0.065}{12}\right)}{1-\left(1+\dfrac{0.065}{12}\right)^{-12\times15}} = \$87.11.$

The payments for the other two loans are calculated similarly; you should get \$116.30, \$148.38, respectively.

b. For the first loan, you'll pay $\$87.11\times12\times15=\$15{,}679.80$ over its term. The total payments for the other loans are computed in a similar fashion – you should get \$27,912 and \$17,805, respectively. Add them together to get \$61,397.

c. The monthly payment for the consolidated loan will be $PMT = \dfrac{\$37{,}500\left(\dfrac{0.065}{12}\right)}{1-\left(1+\dfrac{0.065}{12}\right)^{-12\times15}} = \$326.67$, and the total payment over 15 years will be \$58,801. You'll pay about \$2596 less for the consolidated loan ($\$61{,}397-\$58{,}801=\$2596$), and your monthly payments will be lower for the first ten years (\$326.67 versus $\$87.11+\$116.30+\$148.38=\$351.79$ but higher once the ten year loan is paid off. It would probably be worth it to consolidate the loans. It also means you'll have to keep track of only one loan instead of three, though with automatic withdrawals offered by most banks, this isn't as much of an issue today as it might have been 15 years ago.

53. a. The total fee is $\$500 \times \dfrac{\$20}{\$100 \text{ borrowed}} = \$100.$

b. The interest rate is $\dfrac{\$100}{\$500} = 0.20 = 20\%.$

c. The equivalent APR is $\dfrac{20\%}{2 \text{ weeks}} \times \dfrac{52 \text{ weeks}}{1 \text{ year}} = \dfrac{520\%}{1 \text{ year}}.$

55. a. The annual $(n=1)$ payment is \$8718.46; the monthly $(n=12)$ payment is \$716.43, the bi-weekly $(n=26)$ payment is \$330.43, and the weekly $(n=52)$ payment is \$165.17.

The weekly payment is computed using $PMT = \dfrac{\$100{,}000\left(\dfrac{0.06}{52}\right)}{1-\left(1+\dfrac{0.06}{52}\right)^{-52\times20}} = \$165.17;$ the others are found in a similar fashion.

b. The total payout for each scenario is as follows: \$174,369.20 $(n=1)$; \$171,943.20 $(n=12)$; \$171,823.60 $(n=26)$; and \$171,776.80. The total payout for weekly payments $(n=52)$ is found using

$\dfrac{\$165.17}{\text{week}} \times \dfrac{52 \text{ weeks}}{\text{year}} \times 20 \text{ years} = \$171{,}776.80;$ the other total payments are found in a similar fashion.

c. The total payout decreases as n increases.

TECHNOLOGY EXERCISES

65. a.

Month	Interest	Principal	Balance
			\$7500.00
1	\$56.25	\$38.76	\$7461.24
2	\$55.96	\$39.05	\$7422.19
3	\$55.67	\$39.34	\$7382.85
4	\$55.37	\$39.64	\$7343.21
5	\$55.07	\$39.94	\$7303.27
6	\$54.77	\$40.24	\$7263.04
7	\$54.47	\$40.54	\$7222.50
8	\$54.17	\$40.84	\$7181.66
9	\$53.86	\$41.15	\$7140.51
10	\$53.55	\$41.46	\$7099.05

b. The interest paid in the first month of the loan is \$56.25 and \$38.76 is paid towards the principle.

c. The interest paid in the last month of the loan is \$0.70 and \$94.31 is paid towards the principle.

UNIT 4E: INCOME TAXES

QUICK QUIZ

1. **a.** Gross income is defined as the total of all income you receive.

2. **c.** The first portion of your income is taxed at 10% (the portion being determined by your filing status – see Table 4.9), the next at 15%, and only the last portion is taxed at 25%.

3. **a.** A tax credit reduces your tax bill by the dollar amount of the credit, so your tax bill will be reduced by \$1000.

4. **b.** A tax deduction reduces your taxable income, so if you have a deduction of \$1000, and you are in the 15% tax bracket, your bill will be reduced by 15% of \$1000, or \$150. This answer assumes you are far enough into the 15% tax bracket so that reducing your taxable income by \$1000 means you remain in the 15% bracket.

5. **b.** You can claim deductions of $\$5000 + \$2000 = \$7000$ as long as you itemize deductions.

6. **a.** If you chose to itemize your deductions, the only thing you can claim is the $1000 contribution, and as this is less than your standard deduction of $6100, it won't give relief to your tax bill (in fact, you would pay more tax it you were to itemize deductions in this situation – take the standard deduction).

7. **c.** FICA taxes are taxes levied on income from wages (and tips) to fund the Social Security and Medicare programs.

8. **a.** Joe will pay 7.65% of his income for FICA taxes. Kim pays 7.65% of only her first $127,200; she pays 1.45% of her remaining salary, for a total FICA tax of $0.0765(\$127{,}200) + 0.0145(\$175{,}000 - \$127{,}200) = \$10{,}426.90$, which is 5.96% of her income. David pays nothing at all in FICA taxes because income from capital gains is not subject to FICA taxes.

9. **b.** Assuming all are of the same filing status, Jerome pays the most because his FICA taxes will be highest, followed by Jenny, and then Jacqueline.

10. **c.** Taxes on money deposited into tax-deferred accounts are *deferred* until a later date: you don't have to pay taxes on that money now, but you will when you withdraw the money in later years.

REVIEW QUESTIONS

1. Gross income is all your income for the year, including wages, tips, profits from a business, interest or dividends from investments, and any other income you receive. Contributions to individual retirement accounts (IRAs) and other tax-deferred savings plans, are not taxed. These untaxed portions of gross income are called adjustments and are subtracted from your gross income to determine your adjusted gross income. Most people are entitled to certain exemptions and deductions, which are subtracted from adjusted gross income before taxes are calculated. Once you subtract your exemptions and deductions, you are left with your taxable income. From this, your total tax is calculated. Finally, depending on any tax payments or withholding, you determine if you owe more taxes or receive a refund.

3. Exemptions are a fixed amount per person that you can claim for yourself and each of your dependents to reduce your taxable income. Deductions vary from one person to another. The most common deductions include interest on home mortgages, contributions to charities, and taxes you've paid to other agencies (such as state income taxes and local property taxes). You itemize your deductions if the are greater, in total, than the standard deduction.

5. A tax credit reduces your total tax bill by the amount of the credit while a tax deduction reduces your taxable income by the amount of the credit. A tax reduction only saves you the percentage of the deduction from your highest marginal tax bracket.

7. Taxes collected under FICA are used to pay Social Security and Medicare benefits, primarily to people who are retired. FICA taxes are calculated on all wages, tips, and self-employment income. FICA does not apply to income from sources such as interest, dividends, or profits from sales of stock.

9. Contributions to tax-deferred savings plans count as adjustments to your present gross income and are not part of your taxable income. All earnings in tax-deferred savings plans are also tax deferred.

DOES IT MAKE SENSE?

11. Does not make sense. We know nothing of the rest of the picture for these two individuals – they may have very different deductions, and thus different tax bills.

13. Makes sense. You can deduct interest paid on the mortgage of a home if you itemize deductions, and for many people, the mortgage interest deduction is the largest of all deductions.

15. Makes sense (in some scenarios). Bob and Sue might be hit with the "marriage penalty" when they file taxes, and if they postpone their wedding until the new year, they could avoid the penalty for at least the previous tax year.

17. Makes sense, because at $10,000 your personal exemption and standard deduction would give you $0 taxable income, but being self-employed means you are still subject to the 15.3% self-employment FICA tax.

BASIC SKILLS AND CONCEPTS

19. Antonio's gross income was $\$47,200+\$2400=\$49,600$. His AGI was $\$49,600-\$3500=\$46,100$. His taxable income was $\$46,100-\$4050-\$6350=\$35,700$.

21. Isabella's gross income was $\$88,750+\$4900=\$93,650$. Her AGI was $\$93,650-\$6200=\$87,450$. Her taxable income was $\$87,450-\$4050-\$9050=\$74,350$.

23. Your itemized deductions total to $\$8600+\$2700+\$645=\$11,945$, and since this is smaller than the standard deduction of $12,700, you should claim the standard deduction and not itemize your deductions.

25. Suzanne's gross income is $\$33,200+\$350=\$33,550$, her AGI is $\$33,550-\$500=\$33,050$, and her taxable income is $33,050 - $4050 - $6350 = $22,650. She should claim the standard deduction because it is much higher than her itemized deduction of $450.

27. Wanda's gross income was $\$33,400+\$500=\$33,900$, her AGI was the same, and her taxable income was $\$33,900-\$4050\times 3-\$6350=\$15,400$. She should claim the standard deduction because it is higher than her itemized deduction of $1500.

29. Gene owes $10\%\times(\$9325)+15\%\times(\$35,400-\$9325)=\$4843.75$ (Taxes are rounded to the nearest dollar, so the tax owed is $4844.)

31. Bobbi owes $10\%\times(\$9325)+15\%(\$37,950-\$9325)+25\%\times(\$76,550-\$37,950)$ $+28\%\times(\$77,300-\$76,550)=\$15,086.25$. (Taxes are rounded to the nearest dollar, so the tax owed is $15,086.)

33. Paul owes $10\%\times(\$13,350)+15\%\times(\$50,800-\$13,350)+25\%\times(\$89,300-\$50,800)-\$1000=\$15,577.5$. (Taxes are rounded to the nearest dollar, so the tax owed is $15,578.)

35. Winona and Jim owe $10\%\times(\$18,650)+15\%\times(\$75,900-\$18,650)+25\%\times(\$105,500-\$75,900)-\$2000$ $=\$15,852.5$. (Taxes are rounded to the nearest dollar, so the tax owed is $15,853.)

37. Their bill will be reduced by $500.

39. Her taxes will not be affected because she is claiming the standard deduction, and a $1000 charitable contribution must be itemized.

41. His tax bill will be reduced by 28% of $1000, or $280, provided he is far enough into the 28% bracket that his $1000 contribution does not drop him into the lower bracket.

43. The $1800 per month of your payment that goes toward interest would be deductible, which means you would save $33\%\times\$1800=\$594$ each month in taxes. Thus your $2000 mortgage payment would effectively be only $\$2000-\$594=\$1406$, which is less than your rent payment of $1600. Buying the house is cheaper. This solution assumes that the $\$1800\times 12=\$21,600$ interest deduction would not drop you into the 28% bracket (if it did, the solution would be more complicated).

45. Maria saves 33% of $10,000, or $3300 in taxes, which means the true cost of her mortgage interest is $6700. Steve saves 15% of $10,000, or $1500 in taxes, which means the true cost of his mortgage interest is $8500. This solution assumes their deductions do not drop them into the lower tax bracket. Steve would save less if he took the standard deduction, as it is smaller than the interest he deducted.

47. Luis will pay $7.65\%\times\$28,000=\$2142$ in FICA taxes. His taxable income is $\$28,000-\$4050-\$6350$ $=\$17,600$. He will pay $10\%\times(\$9325)+15\%\times(\$17,600-\$9325)=\$2174$ in income taxes, and thus his total tax bill will be $\$2174+\$2142=\$4316$. This is $\dfrac{\$4316}{\$28,000}=15.4\%$ of his gross income, so his effective tax rate is 15.4%.

49. Jack will pay $7.65\% \times \$44,800 = \$3427$ in FICA taxes. His gross income is $\$44,800 + \$1250 = \$46,050$, and his taxable income is $\$46,050 - \$4050 - \$6350 = \$35,650$. He will pay $10\% \times (\$9325)$ $+ 15\% \times (\$35,650 - \$9325) = \$4881$ in income taxes, and thus his total tax bill will be $\$3427 + \$4881 = \$8308$. This is $\frac{\$8308}{\$46,050} = 18.0\%$ of his gross income, so his effective tax rate is 18.0%.

51. Brittany will pay $7.65\% \times \$48,200 = \$3687$ in FICA taxes. Her gross income is \$48,200. Her taxable income is $\$48,200 - \$4050 - \$6350 = \$37,800$. She will pay $10\% \times (\$9325) + 15\% \times (\$37,800 - \$9325) = \$5204$. in income taxes, and thus her total tax bill will be $\$3687 + \$5204 = \$8891$. This is $\frac{\$8891}{\$48,200} = 18.4\%$ of her gross income, so her effective tax rate is 18.4%.

53. Pierre will pay $7.65\% \times \$120,000 = \$9180$ in FICA taxes, and because his taxable income is $\$120,000 - \$4050 - \$6350 = \$109,600$, he will pay $10\% \times (\$9325) + 15\% \times (\$37,950 - \$9325)$ $+ 25\% \times (\$91,900 - \$37,950) + 28\% \times (\$109,600 - \$91,900) = \$23,670$ in income taxes. His total tax bill is $\$9180 + \$23,670 = \$32,850$ so his overall tax rate is $\frac{\$32,850}{\$120,000} = 27\%$. Like Pierre, Katarina's taxable income is \$109,600, but she is taxed at the special rates for dividends and long-term capital gains. She will pay nothing in FICA taxes, and her income taxes will be $0\% \times (\$37,950) + 15\% \times (\$109,600 - \$37,950) = \$10,748$. Her overall tax rate is $\frac{\$10,748}{\$120,000} = 9.0\%$.

55. Because you are in the 15% tax bracket, every time you make a \$400 contribution to your tax-deferred savings plan, you save $15\% \times \$400 = \$60$ in taxes. Thus your take-home pay is reduced not by \$400, but by \$340.

57. Because you are in the 25% tax bracket, every time you make an \$800 contribution to your tax-deferred savings plan, you save $25\% \times \$800 = \$200$ in taxes. Thus your take-home pay is reduced not by \$800, but by \$600.

FURTHER APPLICATIONS

59. Gabriella's taxable income is $\$96,400 - \$4050 - \$6350 = \$86,000$, so her income tax is $10\% \times (\$9325)$ $+ 15\% \times (\$37,950 - \$9325) + 25\% \times (\$86,000 - \$37,950) = \$17,239$. Roberto's taxable income is $\$82,600 - \$4050 - \$6350 = \$72,200$, so his income tax is $10\% \times (\$9325) + 15\% \times (\$37,950 - \$9325)$ $+ 25\% \times (\$72,200 - \$37,950) = \$13,789$. If they delay their marriage, they will pay $\$17,239 + 13,789$ $= \$31,028$ in taxes. If they marry, their combined AGI will be $\$96,400 + \$82,600 = \$179,000$, and their taxable income will be $\$179,000 - \$4050 \times 2 - \$12,700 = \$158,200$. Their income taxes will be $10\% \times (\$18,650) + 15\% \times (\$75,900 - \$18,650) + 25\% \times (\$153,100 - \$75,900) + 28\% \times (\$158,200 - \$153,100)$ $= \$31,181$, which is slightly higher than the taxes they would pay as individuals, and thus they will face a marriage penalty.

61. Steve's taxable income is $\$185,000 - \$4050 - \$6350 = \$174,600$, so his income tax is $10\% \times (\$9325)$ $+ 15\% \times (\$37,950 - \$9325) + 25\% \times (\$91,900 - \$37,950) + 28\% \times (\$174,600 - \$91,900) = \$41,870$. Mia's tax is, of course, the same, so if they delay their marriage, they will pay $\$\ 41,870 \times 2 = \$83,740$. If they marry, their combined AGI will be $\$\ 185,000 \times 2 = \$370,000$, and their taxable income will be $\$370,000 - \$4050 \times 2 - \$12,700 = \$349,200$. Their income taxes will be $10\% \times (\$18,650)$ $+ 15\% \times (\$75,900 - \$18,650) + 25\% \times (\$153,100 - \$75,900) + 28\% \times (\$233,350 - \$153,100)$ $+ 33\% \times (\$349,200 - \$233,350) = \$90,453$, which is more than the taxes they would pay as individuals, and thus they will face a marriage penalty.

63. a. Deirdre will pay $7.65\% \times (\$90{,}000) = \$6885$ in FICA taxes. Robert will also pay \$6885 in FICA taxes. Jessica and Frank will pay $2 \times (\$6885) = \$13{,}770$ in FICA taxes.

b. From Example 3, Deirdre will pay $\$15{,}639 + \$6885 = \$22{,}524$, Robert will pay $\$11{,}378 + \$6885 = \$18{,}263$, and Jessica and Frank will pay $\$31{,}460 + \$13{,}770 = \$45{,}230$.

c. Deirdre's overall tax rate: $\dfrac{\$22{,}524}{\$90{,}000} = 25.0\%$ Robert's overall tax rate: $\dfrac{\$18{,}263}{\$90{,}000} = 20.3\%$

Jessica and Frank's overall tax rate: $\dfrac{\$45{,}230}{\$180{,}000} = 25.1\%$

d. Serena had the lowest overall tax rate, following by Robert, Deirdre, and Jessica and Frank, in ascending order.

e. Serena receives by far the greatest tax break because all of her income is from investments.

65. a. $24.3\% + 20.4\% + 17.6\% = 62.3\%$ of federal tax returns were filed by this group. $0.1\% + 1.4\% + 4.1\% = 5.6\%$ of total federal income taxes were paid by this group.

b. 2.7% of federal tax returns were filed by this group. 51.6% of total federal income taxes were paid by this group.

c. Answers will vary.

d. Answers will vary.

UNIT 4F: UNDERSTANDING THE FEDERAL BUDGET

QUICK QUIZ

1. b. Bigprofit.com had more outlays (expenses) than receipts (income) by \$1 million, so it ran a deficit for 2009 of \$1 million. Its debt is the total amount owed to lenders over the years; this is \$7 million.
2. c. The federal debt is about \$20 trillion dollars, and when divided evenly among all 325 million U.S. citizens, it comes to about \$62,000 per person.
3. a. Tax revenues are about average.
4. b. Discretionary outlays differ from mandatory outlays in that Congress passes a budget every year that spells out where the government will spend its discretionary funds.
5. a. Interest on the debt must be paid so that the government does not default on its loans, and Medicare is an entitlement program that is part of mandatory spending. On the other hand, all money spent on defense is considered discretionary spending.
6. a. Mandatory expenses include money spent on Social Security, Medicare, interest on the debt, government pensions, and Medicaid, and together these programs constitute almost 60% of the budget (see Figure 4.14).
7. b. The \$50 billion is supposed to be deposited into an account used to pay off future Social Security recipients, but instead, the government spends the money at hand, and essentially writes IOUs to itself, and places these in the account.
8. b. Publicly held debt is the money the government owes to those who have purchased Treasury bills, notes, and bonds.
9. b. The gross debt is the sum of all of the money that the government owes to individuals who have purchased bonds, and the money it owes to itself.
10. c. The amount of money set aside for education grants is miniscule in comparison to the billions of dollars the government will be required to spend on Social Security in 2036.

REVIEW QUESTIONS

1. Receipts, or income, represent money that has been collected. Outlays, or expenses, represent money that has been spent. Net income = receipts − outlays. If net income is positive, the budget has a surplus. If net income is negative, the budget has a deficit.

3. Years of running deficits causes the federal debt to continues to grow, meaning that the mandatory interest payments are virtually certain to rise. This, along with the other mandatory expenses, means that federal outlays will continue to rise. Since politicians have found it very difficult to reduce these expenses and federal receipts are difficult to increase, it because evermore difficult to balance the federal budget.

5. For receipts, over 80% of the federal government's receipts come from the combination of individual income taxes and the FICA taxes collected primarily for Social Security and Medicare. The remainder come from corporate income taxes, excise taxes (including taxes on alcohol, tobacco, and gasoline), and a variety of "other" categories that include gift taxes and fines collected by the government. Mandatory outlays are expenses that are paid automatically unless Congress acts to change them. Discretionary outlays are the ones that Congress must vote on each year and that the President must then sign into law.

7. The Social Security trust fund exists to ensure money will be available when needed for future retirees. At the present time, there is no actual money in the Social Security trust fund. If this continues, the government may not have enough money to pay retirees in the future.

DOES IT MAKE SENSE?

9. Makes sense. Each U.S. citizen would have to spend about $62,000 to retire the federal debt, and it is certainly true that this exceeds the value of most new cars.

11. Does not make sense. The financial health of the government is very much dependent upon what happens with Social Security and the FICA taxes collected to fund it. The term "off-budget" is nothing more than a label that separates the Social Security program from the rest of the budget.

13. Does not make sense. There is no guarantee the government will have the funds to pay for social security 40 years from now.

BASIC SKILLS AND CONCEPTS

15. a. Your receipts are $38,000, and your outlays are $\$12,000+\$6000+\$1200+\$8500=\$27,700$, so you have a surplus.

b. A 3% raise means your receipts will be $\$38,000+3\%\times\$38,000=\$39,140$, and your outlays will be $\$27,700+\$8500=\$36,200$, so you'll still have a surplus, though it will be smaller.

c. Your receipts will be $39,140 (see part (b)), and your outlays will be $\$27,700\times1.01+\$7500=\$35,477$, so you will be able to afford it.

17. a. The interest on the debt will be $8.2\%\times\$773,000=\$63,000$ (rounded to the nearest thousand dollars).

b. The total outlays are $\$600,000+\$200,000+\$250,000+\$63,000=\$1,113,000$, so the year-end deficit is $\$1,050,000-\$1,113,000=-\$63,000$, and the year-end accumulated debt is $-\$773,000-\$63,000$ $=-\$836,000$.

c. The interest on the debt in 2019 will be $8.2\%\times\$836,000=\$69,000$ (rounded to the nearest thousand dollars).

d. The total outlays are $\$600,000+\$200,000+\$69,000=\$869,000$, so the year-end surplus is $\$1,100,000-\$869,000=\$231,000$, and the year-end accumulated debt is $-\$836,000+\$231,000$ $=-\$605,000$, assuming all surplus is devoted to paying down the debt.

e. Answers will vary.

19. $$\frac{\$20\times10^{12}}{160\times10^{6}\text{ workers}}=\frac{\$125,000}{\text{worker}}$$

21. 2000 Surplus: $\dfrac{\$236\times10^{9}}{\$10,100\times10^{9}}=2.3\%$ 2010 Deficit: $\dfrac{\$1294\times10^{9}}{\$14,800\times10^{9}}=8.7\%$

2000 Debt: $\dfrac{\$5600\times10^{9}}{\$10,100\times10^{9}}=55.4\%$ 2010 Debt: $\dfrac{\$13,600\times10^{9}}{\$14,800\times10^{9}}=91.9\%$

23. 2010 Deficit: $\dfrac{\$1294\times10^9}{\$14,800\times10^9}=8.7\%$ 2020 Debt: $\dfrac{\$22,500\times10^9}{\$21,900\times10^9}=102.7\%$

2020 Deficit: $\dfrac{\$540\times10^9}{\$21,900\times10^9}=2.5\%$ This is a change of $\dfrac{102.7\%-91.9\%}{91.9\%}\approx11.8\%$.

This is a change of $\dfrac{2.5\%-8.7\%}{8.7\%}\approx-71.3\%$.

25. At 1.7%, the interest payment would be $1.7\%\times\$22.5\times10^{12}=3.825\times10^{11}$, or about \$383 billion. With the 0.5% increase, the interest payment would increase to $2.2\%\times\$22.5\times10^{12}=4.95\times10^{11}$, or \$495 billion, , which is a $\dfrac{2.2\%-1.7\%}{1.7\%}\approx29.4\%$ increase.

27. Revenue from individual income taxes was $49\%\times\$3.31\times10^{12}=\$1.62\times10^{12}$, or \$1.62 trillion. Revenue from social insurance was $33\%\times\$3.31\times10^{12}=\$1.09\times10^{12}$, or \$1.09 trillion.

29. Spending would increase by $1.6\%\times\$3.89\times10^{12}=\$6.22\times10^{10}$, or \$62.2 billion, while revenue would increase by $2\%\times\$3.31\times10^{12}=\$6.62\times10^{10}$, or \$66.2 billion. The deficit would decrease, since overall, revenue is less than spending.

31. a. Spending on Social Security and Medicare was $23\%+15\%=38\%$ of spending, or $38\%\times\$3.89\times10^{12}=\$1.48\times10^{12}$, or \$1.48 trillion.

b. Discretionary spending was $16\%+15\%=31\%$ of spending, or $31\%\times\$3.89\times10^{12}=\$1.21\times10^{12}$, or \$1.21 trillion.

c. Spending would decrease by $2\%\times\$3.89\times10^{12}=\$7.78\times10^{10}$, or \$77.8 billion, which is $\dfrac{\$7.78\times10^{10}}{\$1.21\times10^{12}}\approx6.4\%$.

33. \$40 billion − \$180 billion = −\$140 billion; In other words, despite the fact that the government proclaimed a surplus of \$40 billion for this year, it actually ran a deficit of \$140 billion due to the fact that the money earmarked for the Social Security trust fund was never deposited there (it was spent on other programs).

35. The government could cut discretionary spending, it could borrow money by issuing Treasury notes, or it could raise taxes (of course, a combination of these three would also be an option).

FURTHER APPLICATIONS

37. It would take about $\$20\times10^{12}\times\dfrac{1\text{ s}}{\$1}\times\dfrac{1\text{ hr}}{3600\text{ s}}\times\dfrac{1\text{ d}}{24\text{ hr}}\times\dfrac{1\text{ yr}}{365\text{ d}}\approx634{,}000$ years.

39. In ten years, the debt will be $\$20\times10^{12}(1+0.01)^{10}=\$2.21\times10^{13}$, or about \$22.1 trillion. In 50 years, it will be $\$20\times10^{12}(1+0.01)^{50}=\$3.29\times10^{13}$, or about \$32.9 trillion.

41. $$PMT=\frac{\$14.5\times10^{12}\left(\frac{0.03}{1}\right)}{1-\left(1+\frac{0.03}{1}\right)^{-1\times10}}=\$1.70\times10^{12}$$, which is \$1.70 trillion per year.

43. $\$20\times10^{12}\times\dfrac{1\text{ wk}}{\$163\times10^6}\times\dfrac{1\text{ yr}}{52\text{ wk}}\approx2360$ years.

UNIT 5A: FUNDAMENTALS OF STATISTICS

QUICK QUIZ

1. **a.** The population is the complete set of people or things that are being studied.
2. **c.** Those who donated money to the governor's campaign are much more likely to approve of his job.
3. **a.** A representative sample is one where the characteristics of the sample match those of the population.
4. **c.** Each city is considered a different strata.
5. **b.** Those who do not receive the treatment (in this case, a cash incentive) are in the control group. Note that the teacher could also randomly select a third group of students who do not even know they are part of the experiment – she could study their attendance rates, and this would be another control group.
6. **c.** The experiment is not blind, because the participants know whether they are receiving money or not, and the teacher knows to which group each student belongs.
7. **a.** The purpose of a placebo is to control for psychological effects that go along with being a participant of the study.
8. **c.** A single-blind experiment is one where the participants do not know whether they are part of the treatment group or the control group, but the person conducting the experiment does know.
9. **b.** With a margin of error of 3%, both Poll X and Poll Y have predictions that overlap (in that Poll X predicts Powell will receive 46% to 52% of the vote, and Poll Y predicts he will receive 49% to 55% of the vote), and so they are consistent in their predictions.
10. **b.** The confidence interval is simply the results of the poll plus-or-minus the margin of error.

REVIEW QUESTIONS

1. Statistics is the science of collecting, organizing, and interpreting data. Statistics are also the data (numbers or other pieces of information) that describe or summarize something.

3. The five basic steps in a statistical study, which can be used to draw conclusions about a larger population, are:

   1. State the goals of your study. That is, determine the population you want to study and exactly what you would like to learn about it.

   2. Choose a representative sample from the population.

   3. Collect raw data from the sample and summarize these data by finding sample statistics of interest.

   4. Use the sample statistics to infer the population parameters.

   5. Draw conclusions: Determine what you learned and how it addresses your goals.

   As an example, Nielsen uses this process to generalize what shows most of the population is watching.

5. Examples will vary. A statistical study suffers from bias if its design or conduct tends to favor certain results.

7. A placebo lacks the active ingredients of the treatment being tested in a study, but looks or feels enough like the treatment so that participants cannot distinguish whether they are receiving the placebo or the real treatment. The placebo effect is a situation where patients improve simply because they believe they are receiving a useful treatment. An experiment is single-blind if the participants do not know whether they are members of the treatment group or members of the control group, but the experimenters do know. An experiment is double-blind if neither the participants nor the experimenters (people administering the treatment) know who belongs to the treatment group and who belongs to the control group. Blinding helps to mitigate the placebo effect.

DOES IT MAKE SENSE?

9. Does not make sense. A sample is a subset of the population, and thus it cannot exceed the size of the population.

11. Does not make sense. The control group should no treatment, not a different treatment than that given to the treatment group.

13. Does not make sense. The only way to conduct a survey with a margin of error of zero is to get a response from every member of the population.

BASIC SKILLS AND CONCEPTS

15. The population is all Americans, and the sample was the set of 1001 Americans surveyed. The population parameters are the opinions of Americans on Iran, and the sample statistics consist of the opinions of those who were surveyed.

17. The population is the set of stars in the galaxy, and the sample is the set of stars selected for measurement. The population parameter is the average distance from Earth of all of the stars in the galaxy, whereas the sample statistic is the average distance from Earth of just those stars in the sample.

19. The population is the set of all senior executives, and the sample is the set of 150 senior executives surveyed. The population parameters is the most common job interview mistake according to all senior executives, and the sample statistic is the most common job interview mistake according to the 150 executives in the sample; "little or no knowledge of the company."

21. *Step* 1: The goal is to determine the percentage of high school seniors that regularly use a cell phone while driving among a population of high school seniors. *Step* 2: Randomly select (from a list of all students at the school) a sample from the population. *Step* 3: Determine the percentage of those in the sample who claim to use a cell phone regularly while driving., either with an interview, a survey, or a device that measures more precisely when the cell phone is in use. *Step* 4: Infer the percentage of all seniors who use a cell phone regularly while driving. *Step* 5: Assess the results and formulate a conclusion.

23. *Step* 1: The population of this study is American college students, and the goal is to determine the percentage of the population that attend home basketball games. *Step* 2: Choose a representative sample of college students. *Step* 3: Survey students and determine the percentage of those in the sample that attend home basketball games. *Step* 4: Infer the percentage of the population who attend home basketball games. *Step* 5: Assess the results and draw conclusions.

25. *Step* 1: The population in the study is all golden retrievers, and the goal is to determine the average lifetime of all golden retrievers. *Step* 2: Select a representative sample of golden retrievers, possibly from records of randomly selected veterinarians. *Step* 3: Determine the average lifetime of golden retrievers in the sample. *Step* 4: Infer the average lifetime all golden retrievers. *Step* 5: Assess the results and formulate a conclusion.

27. The first 100 first-year students met in the student union is the most representative sample for first-year students at a particular school because they are a more random sample of students at that school. The other samples are not likely to be representative since they would be biased towards subgroups within the population.

29. This is an example of stratified sampling as the sample of taxpayers is broken in several categories (or strata).

31. This is an example of stratified sampling as the participants are divided into groups based on age.

33. This is a simple random sampling because the people are selected at random.

35. This is an observational study, with a case-control element, where the cases are those who have a tendency to lie, and the controls are those who do not lie.

37. This is a retrospective, observational study, with a case-control component. The cases are runners with a specific type of injury, and the controls are runners without that injury.

39. This is an observational study with no case-control (non-retrospective).

41. An experiment with two treatment groups and a control group would answer this question best. No blinding would be necessary.

43. An retrospective, observational study would be sufficient to determine which National Basketball Association teams with high-altitude home courts have better records. The cases are teams with high-altitude home courts, and the controls are the other teams.

45. An experiment would answer this question best. Recruit people to participate in the study, and ask one group to take a multi-vitamin every day, while the other group receives a placebo. Observe the rate of strokes among each group to determine the effect of multi-vitamins.

47. The confidence interval is $53.0\% - 2.5\% = 50.5\%$ to $53.0\% + 2.5\% = 55.5\%$, and because there are only two candidates, it is likely (though not guaranteed) the Republican will win.

49. The confidence interval for the most recent data shows that $46\% - 3\% = 43\%$ to $46\% + 3\% = 49\%$ of Americans support legalized abortion, so 51% to 57% oppose legalized abortion, so you can claim that a majority of Americans oppose legalized abortion.

FURTHER APPLICATIONS

51. a. The treatment group received the new drug; the control group received the placebo.

    b. Significantly more patients improved with the treatment compared to the placebo. However, fewer than half of the patients in the treatment group improved.

    c. Possibly; nearly 20% of the patients showed improvement with the placebo.

    d. Answers will vary, but might refer to side effects, alternative treatments, and/or cost.

53. a. The population is all Americans. The population parameters are the percentages of men and women who regift.

    b. The sample is 900 registered voters. The sample statistics are the percentages of men and women who regift.

    c. This is an observational study.

    d. Answers will vary. One example is how the sample was selected.

55. a. Population is registered American voters; population parameter is percentage of registered American voters who are concerned about fake news.

    b. Sample is the 1006 registered voters surveyed; sample statistic is percentage of voters in the sample who are concerned about fake news.

    c. This is an observational study.

    d. Answers will vary. One example is how the sample was selected.

57. a. Population is all adult Americans; population parameter is percentage of Americans who approved of the ruling.

    b. Sample is the adults who were surveyed; sample statistic is the percentage of respondents in the sample who approved of the ruling (30.8% in 1974 and 39.1% in 2014).

    c. This is an observational study.

    d. Answers will vary. One example is how the sample was selected.

TECHNOLOGY EXERCISES

67. a. Answers will vary

    b. Answers will vary.

    c. The mean of a large number of random numbers between 0 and 1 should be near 0.5.

UNIT 5B: SHOULD YOU BELIEVE A STATISTICAL STUDY?

QUICK QUIZ

1. **c.** When the wrong technique is used in a statistical study, one should put little faith in the results.

2. **b.** While it is possible that an oil company can carry out legitimate and worthy research on an environmental problem it caused, the opportunity (and temptation) for bias to be introduced should make you wary of the results.

3. **a.** The fact that the researchers interviewed only those living in dormitories means the sample chosen may not be representative of the population of all freshmen, some of whom do not live on campus.

4. **b.** The poll that determines the winner suffers from participation bias, as those who vote choose to do so. Also true is that those who vote almost certainly watch the show, while the vast majority of Americans do not watch it, nor do they care who wins.

5. **b.** The quantities that a statistical study attempts to measure are called variables of interest, and this study is measuring the weights of 6-year-olds.

6. **a.** The quantities that a statistical study attempts to measure are called variables of interest, and this survey is measuring the number of visits to the dentist.

7. **c.** It's reasonable to assume that people who use sunscreen do so because they spend time in the sun, and this can lead to sunburns.

8. **c.** The availability error is best avoided by carefully choosing the order in which answers to a survey are presented, and in this case, it's best to switch the order for half of the people being polled.

9. **b.** A self-selected survey suffers from participation bias, where people make the choice to participate (that is, they select themselves, rather than being randomly selected by a polling company).

10. **b.** Carefully conducted statistical studies give us good information about that which is being studied, but it's always possible for unforeseen problems (such as confounding variables) to arise, and the "95% confidence" that was spoken of in Unit 5A (and expanded upon in Unit 6D) can always come into play.

REVIEW QUESTIONS

1. See the Information Box on page 309.

3. The variables of interest in a statistical study are the items or quantities that the study seeks to measure.

DOES IT MAKE SENSE?

5. Does not make sense. A TV survey with text responses suffers from participation bias, whereas a survey carried out using random sampling will more likely than not yield better results.

7. Does not make sense. Statistical studies can rarely (if ever) make the claim that the results are proven beyond all doubt.

BASIC SKILLS AND CONCEPTS

9. There is no reason to doubt the results of this study based on the information given.

11. A study done by a liberal group to assess a new Republican budget plan could easily include biases, so you should doubt the results.

13. You should doubt a poll that suffers from participation bias, as this call-in poll does.

15. It is very difficult to quantify or even define optimism, and thus this study is essentially meaningless (and you should not trust it, unless you believe optimism was correctly defined and measured).

17. You should question the results of this study because other factors should also be considered.

19. You should doubt the results of this study because it uses self reporting, which is not always accurate (especially in cases of sensitive issues, like alcohol abuse).

21. This is a reasonable claim, especially if the effect of inflation is taken into account.

23. Depending on the reliability of the president's evidence, this could be a reasonable claim. You should look for more research of the subject.

25. The Chamber of Commerce would have no reason to distort its data, so the claim is believable.

FURTHER APPLICATIONS

27. Political bias may have been present. Other estimates of attendance ranged from 250,000 to 900,000.

29. A potential selection bias exists in this survey because the people who shop on Saturday morning may not be a representative sample of the population.

31. The poll organization is reputable and most likely used a random sampling method to avoid selection bias.

33. The poll organization is reputable and most likely used a random sampling method to avoid selection bias.

35. Answers will vary.

37. There is no mention of how the respondents were selected or what questions was asked that had 60% of adults avoiding visits to the dentist because of fear.

39. There is no mention of how the respondents were selected, how many responded, or how the quality of the restaurants was measured.

41. We do not know what question was asked to obtain the responses (did the question just ask for a favorite vegetable, or were potatoes one of a list of options?) We also do not know how the sample was selected.

43. A headline like this implies "illegal drugs" to most readers, and the government study includes smoking and drinking (both of which are legal and largely acceptable practices in many segments of society). The study is based upon top movie rentals, and yet the headline implies all movies. It's a misleading headline.

45. The questions have a different population. The population for the first question is all people who have dated on the Internet. The population for the second question is all married people.

47. a. $\frac{48}{339} = 0.14 = 14\%$

b. $0.1\% \times 38{,}000 = 38$ people

c. The 0.1% error rate assumed in part (b) would account for 38 of the 48 people who said they were noncitizens who voted, so that in fact there were only 10 noncitizen voters in the survey. If the error rate were 0.13%, it would account for all those who claimed to be noncitizen voters.

d. The fact that there were zero voters among the 85 people who gave consistent answers about being noncitizens suggests that no noncitizens voted.

UNIT 5C: STATISTICAL TABLES AND GRAPHS

QUICK QUIZ

1. **b.** The relative frequency of a category expresses its frequency as a fraction or percentage of the total, and thus is $50/200 = 0.25$.

2. **c.** We would need to know the number of A's assigned before being able to compute the cumulative frequency.

3. **b.** Qualitative data describe non-numerical categories.

4. **c.** The sizes of the wedges are determined by the percentage of the data that corresponds to a particular category, and this is the same as the relative frequency for that category.

5. **a.** Depending on how you want to display the data, you could actually use any of the three options listed. However, a line chart is used for numerical data, which means you would have to suppress the names of the tourist attractions (your categories would become bins, such as "0-2 million"). For a pie chart, you would need to compute the relative frequencies of the data, and while that's not terribly difficult, it requires an extra step. It would be most appropriate to use a bar graph, with the various tourist attractions as categories (e.g. Disney World, Yosemite National Park, etc.), and the annual number of visitors plotted on the y-axis.

6. **c.** The ten tourist attractions are the categories for a bar graph, and these belong on the horizontal axis.

7. **a.** With 100 data points, each precise to the nearest 0.001, it would be best to bin the data into several categories.

8. **c.** A line chart is often used to represent time-series data.

9. **c.** A histogram is a bar graph for quantitative data categories.

10. **b.** For each dot in a line chart, the horizontal position is the center of the bin it represents.

REVIEW QUESTIONS

1. A frequency table contains two columns: The first column lists all the categories of data and the second column lists the number of data values (the frequency) in each category. The relative frequency of any category is the fraction (or percentage) of the data values that fall in that category. The cumulative frequency of any category is the number of data values in that category and all preceding categories.

3. Binning allows quantitative data to be sorted into a manageable number of categories. For example, a large set of numbers could be binned into categories of 0–9, 10–19, 20–29, etc.

5. Without proper labels, a graph is meaningless. See the Information Box on page 322 for a description of labels that should be included on graphs.

DOES IT MAKE SENSE?

7. Does not make sense. A frequency table has two columns, one of which lists the frequency of the various categories.

9. Make sense. As long as there are at least 40 students in the class, there could be 40 students with a grade of C or lower.

11. Makes sense. The total of all frequencies in a pie chart should always be 100% (or very near to that – sometimes the total does not add up to 100% due to rounding).

BASIC SKILLS AND CONCEPTS

13.

Grade	Freq.	Rel. freq.	Cum. freq.
A	6	.20	6
B	6	.20	12
C	10	0.33	22
D	5	0.17	27
F	3	0.10	30
Total	30	1.00	30

15. Eye color, which is non-numerical, is qualitative data.

17. Home prices are quantitative data because it is a numerical measurement of money.

19. Flavors of ice cream are qualitative data because they are non-numerical data.

21. Annual salaries are quantitative data because it is a numerical measurement of money.

23.

Bin	Freq.	Rel. freq.	Cum. freq.
95–99	3	0.15	3
90–94	2	0.10	5
85–89	3	0.15	8
80–84	2	0.10	10
75–79	4	0.20	14
70–74	1	0.05	15
65–69	3	0.15	18
60–64	2	0.10	20
Total	20	1.00	20

25.

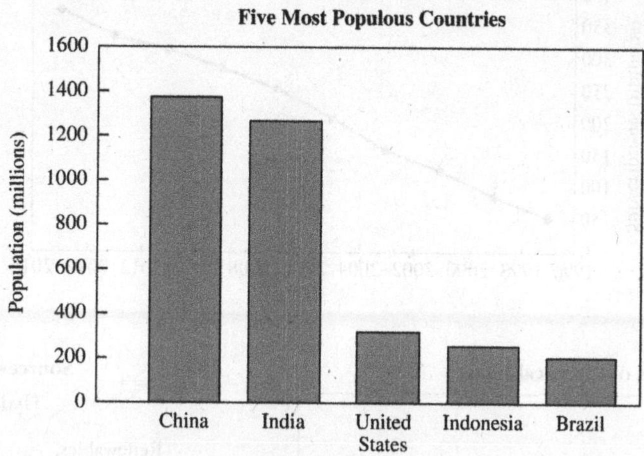
Five Most Populous Countries
Population (millions)
1600
1400
1200
1000
800
600
400
200
0
China
India
United States
Indonesia
Brazil

27.

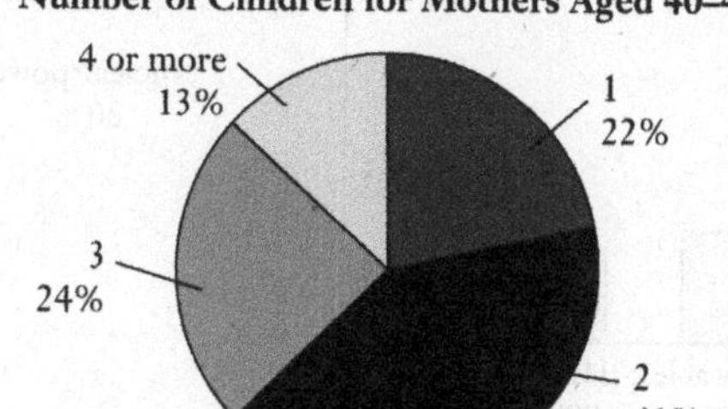
Number of Children for Mothers Aged 40–44
4 or more
13%
1
22%
3
24%
2
41%

29.

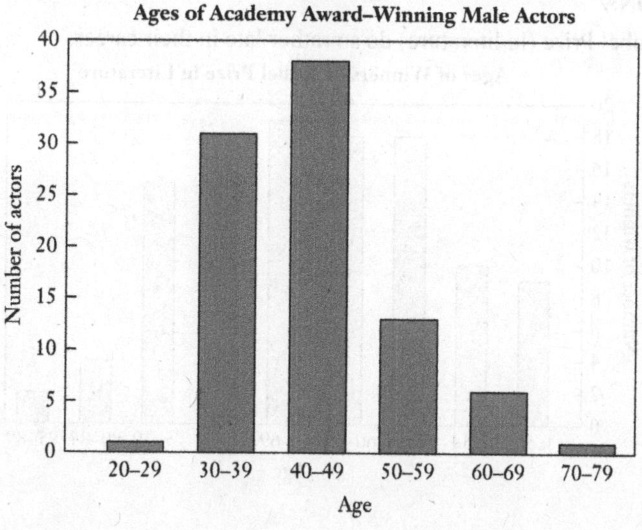
Ages of Academy Award–Winning Male Actors
Number of actors
40
35
30
25
20
15
10
5
0
20–29
30–39
40–49
50–59
60–69
70–79
Age

31. The growth is linear.

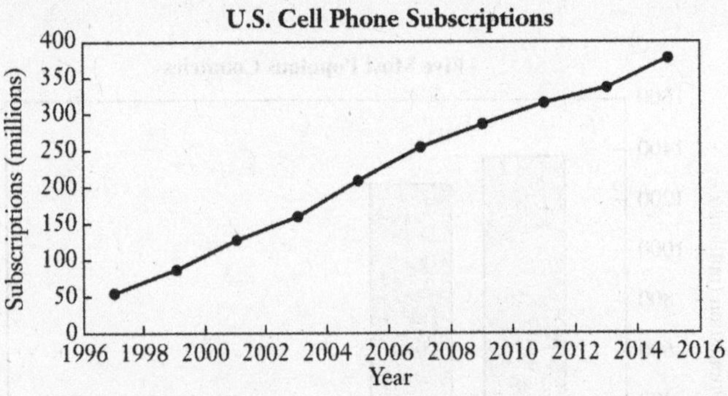

33. Answers will vary.

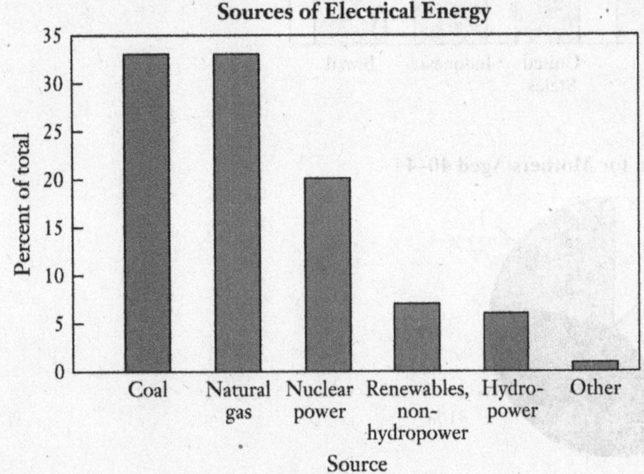

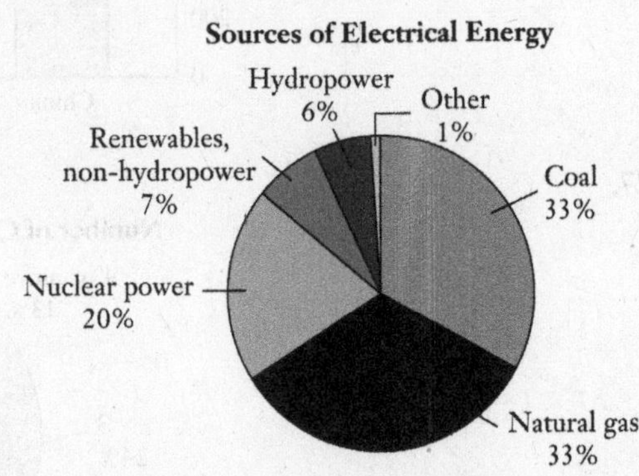

FURTHER APPLICATIONS

35. People who win the Nobel Prize (in literature) do so rather late in their career.

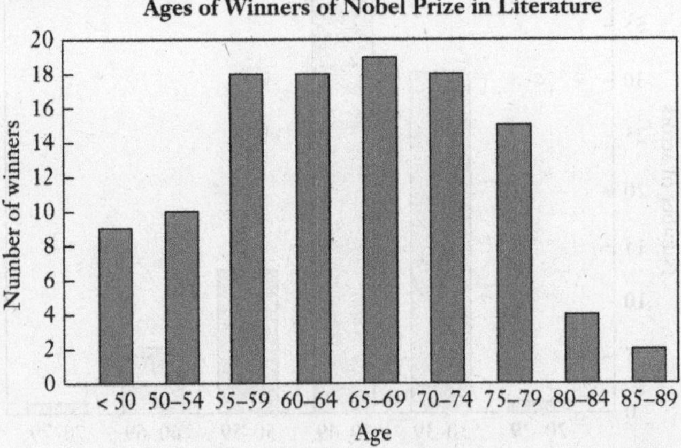

37. Answers will vary.

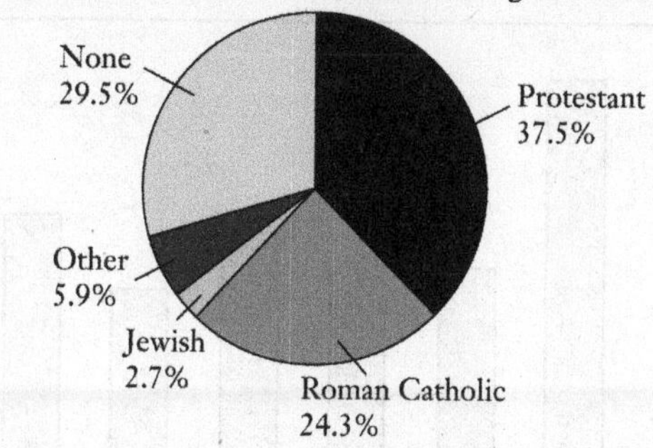

39. The percentage of the U.S. population that was foreign born fell from 1940 to 1970 and has been steadily increasing since that time.

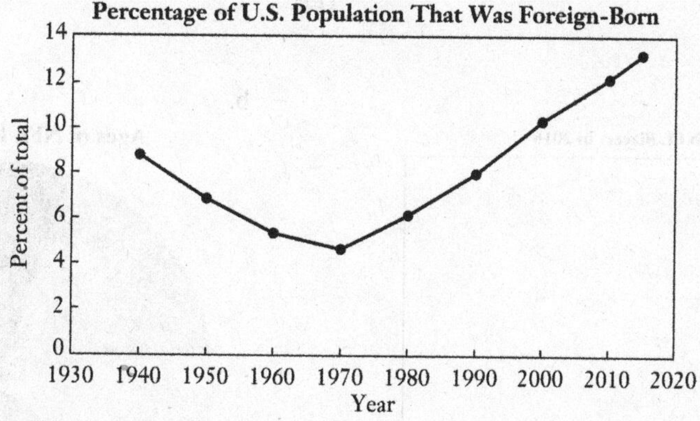

TECHNOLOGY EXERCISES

47. a.

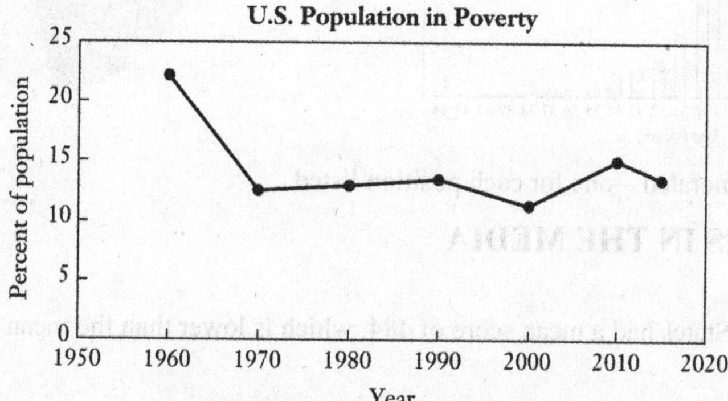

47. (continued)

b.

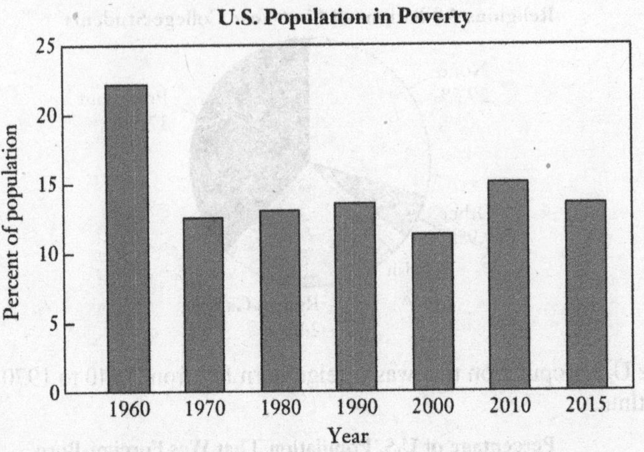

49. a.

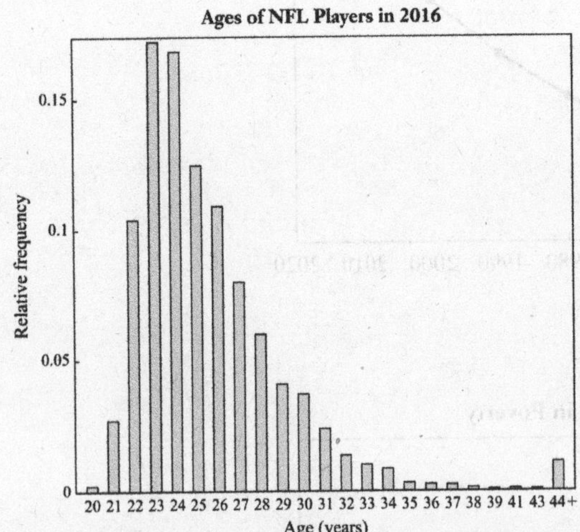

b.

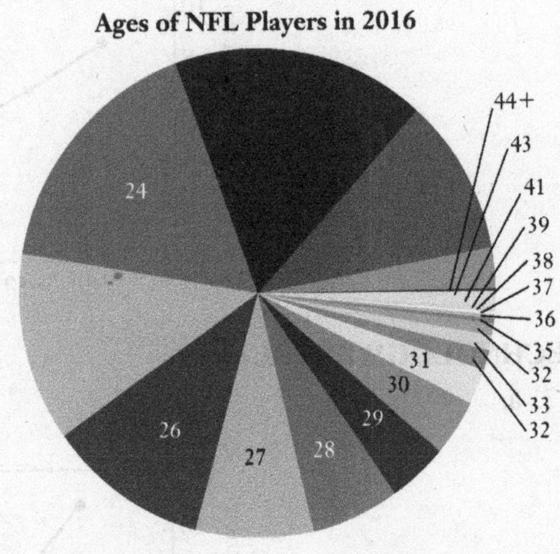

c. 23 pie charts are generated—one for each position listed.

UNIT 5D: GRAPHICS IN THE MEDIA

QUICK QUIZ

1. **c.** Boys in the United States had a mean score of 484, which is lower than the mean score of 513 for girls in Canada.
2. **c.** The red segment in the second bar from the top is farther to the right because the light and dark blue segments representing tuition, fees, and room and board are larger.
3. **a.** The height of the pneumonia wedge at 1940 is about 60.
4. **a.** The color of Missouri corresponds to the 2000 – 2999 gallons of oil equivalent.
5. **c.** Iowa is contained between contour lines that are marked 30°F and 40°F, so the temperature in Des Moines is between these values.
6. **c.** When contour lines are close together, the data they represent change more rapidly compared to areas where the lines are far apart.
7. **a.** This is the reason stated in the description given in the explanation of Figure 5.19.

8. **b.** Because the scale on the *y*-axis is exaggerated, the changes appear to be larger than they really were.

9. **c.** Each tick mark represents another power of ten.

10. **b.** For all of the years shown in the figure, there is a positive percent change, which means the cost of college was increasing; it increases at a smaller rate when the graph is decreasing.

REVIEW QUESTIONS

1. Multiple bar graphs are simple extensions of regular bar graphs. For each category, separate bars are drawn representing each data set. In stack plots, each data set is represented by single bar, where the categories are represented by a segment of the entire bar. Multiple line charts use a different line to represent each data set.

3. Three-dimensional graphics are graphics that have an added element of depth perception. In some cases, the third dimension is only present to create a more appealing image, adding no additional information to the graphic. In other cases, the third dimension adds information, such as an additional axis.

5. If an entire two-dimensional object (where only one dimension represents the value in question) is enlarged, the change in area can give the perception that the value has changed more than the graphic intended to show. Scales than exclude zero can make small-scale trends easier to observe, but can overemphasize changes between different values.

7. A percentage change graph with descending bars (or a descending line) tells us only that value in question was increasing by smaller amounts, not that it was actually decreasing.

DOES IT MAKE SENSE?

9. Does not make sense. Using three-dimensional bars for two-dimensional data is a cosmetic change; it adds no information.

11. Makes sense. Experimenting with different scales on the axes of graphs allows you to highlight certain aspects of the data.

BASIC SKILLS AND CONCEPTS

13. a. Earnings for men with a bachelor's degree are $\frac{\$71,385-\$51,681}{\$51,681}=0.381=38.1\%$ higher than for women with a bachelor's degree.

b. Earnings for men with a professional degree are $\frac{\$131,189-\$82,473}{\$82,473}=0.591=59.1\%$ higher than for women with a professional degree.

15. a. In 45 years, the average male college graduate would earn $45\times\$71,385-45\times\$41,569=\$1,341,720$ more than the average male high school graduate.

b. In 45 years, the average female with a professional degree would earn $45\times\$82,473-45\times\$51,681$ $=\$1,385,640$ more than the average female college graduate.

17. a. Boys in Canada (523) scored lower than girls in South Korea (544).

b. Boys in Finland (517) scored lower than girls in Japan (527).

19. a. Books and supplies vary the least, probably because students at all types of colleges need the same number of books and other supplies on average, and these do not vary much in cost.

b. Transportation, because commuters must spend more money to get to and from school, while students living on campus do not have to pay commuting expenses.

21. a. The death rates for cancer increased, tuberculosis decreased, cardiovascular disease increased, and pneumonia decreased.

b. The death rate was the greatest around 1950, where it was about 510 deaths per 100,000.

c. The death rate for cancer in 2000 was about 193 deaths per 100,000.

d. If the lines of the graph were extended, cancer would overtake cardiovascular disease as the leading cause of death in 2050.

23. a. The thickness of a wedge at a particular time tells you its value at that time. About 5% of the budget went to Medicare in 1976; it rose to about 16% in 2016.

    b. during 1967 – 1969

    c. About 10% of the budget went to Social Security in 1966; it increased to about 25% in 2016.

    d. No, because dollar figures are not given.

25. With a few isolated exceptions, the highest rates occur in the southern part of the country.

27. a. Since points A and B correspond to summits, you would walk downhill from A to C.

    b. Since there are more contour lines between B and D, your elevation would change more between those two points.

29. a. In 2010, the over-65-year-old group constituted 12% of the population.

    b. By 2050, the over-65-year-old group is projected to be to about 20%.

    c. The percentage of 45 to 54-year-olds was about 15% in 2010, and is projected to decrease to about 10% in 2050.

31. a. 290

    b. Yes, the two groups must overlap.

33. No. The height of the barrels represents oil consumption, but the eye is drawn to the volume of the barrels.

35. a. The ratio of refugees from Iraq to refugees from Syria is about $\frac{25,000}{79,000} = 0.32$. The ratio of the lengths of the corresponding bars in the graph is about $\frac{1}{12} = 0.09$.

    b. Because the scale on the y-axis does not start at zero, the ratio appears smaller than it really is.

    c.

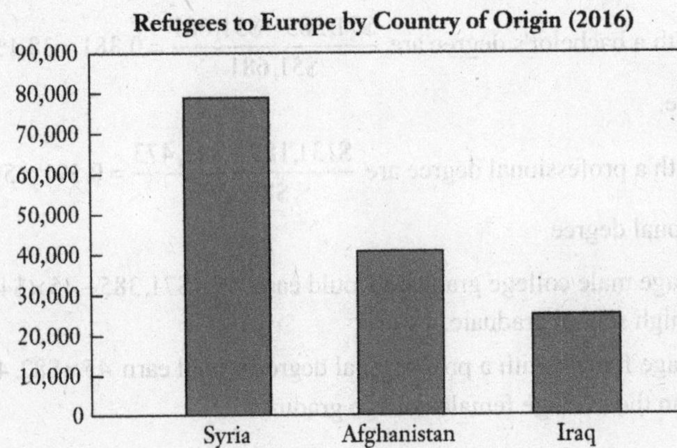

37. a.

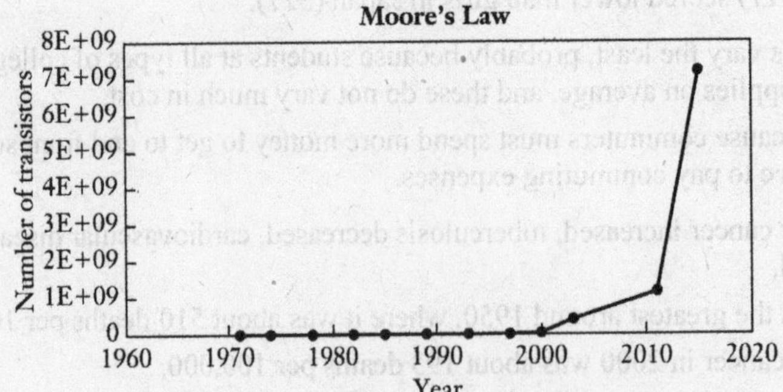

37. (continued)

b.

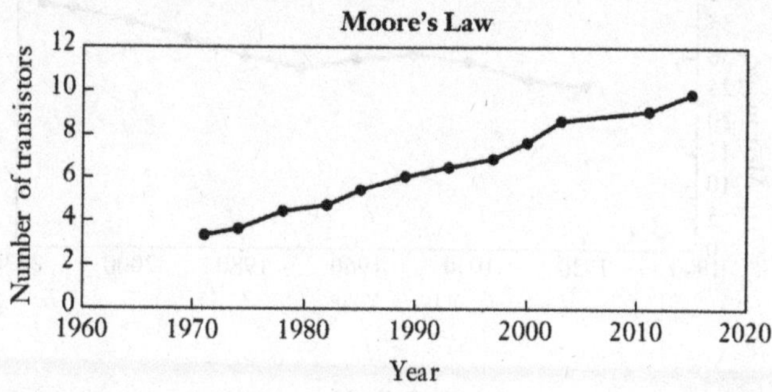

c. The exponential scale clearly shows the rapid gains in computer speeds, while the ordinary scale makes it impossible to see any detail in the early years shown on the graph. More generally, exponential scales are useful whenever data vary over a large range of values.

39. a. Public college costs rose by about 13% in the 2003–04 academic year.

b. The percentage increase in private college costs in the 2003–04 year was about 5%.

c. Private college costs rose more in dollars, because 5% of the 2003–2004 costs for private colleges is more than 13% of the 2003–2004 costs for public colleges.

41. Most of the growth occurred after 1950.

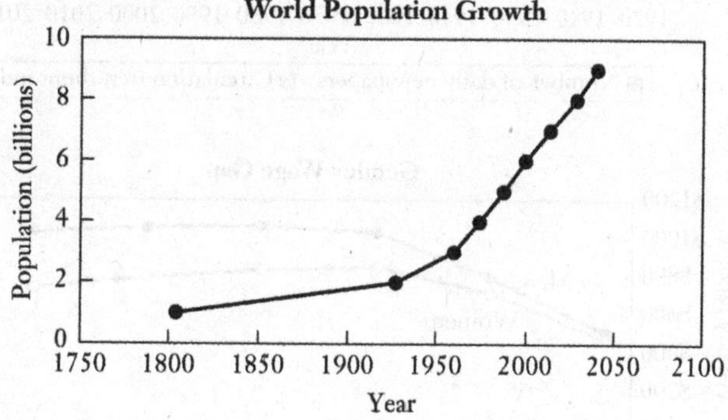

FURTHER APPLICATIONS

43. a. China, India and Russia all had to import grain because their net grain production is negative.

b. China, India, and Russia are all expected to need grain imports because their projected 2030 net grain production is negative.

c. The graph shows that both China and India will require significant grain imports, and thus (based only on the information shown in the graph), it is likely world grain production will need to increase.

45. a. 1957

b. 1990

c. Answers will vary. One possible answer is to show the similarity between the two time periods.

47. Answers will vary.

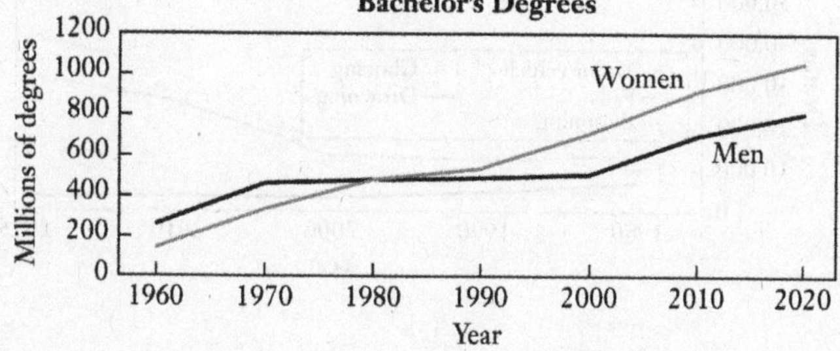

49. Answers will vary.

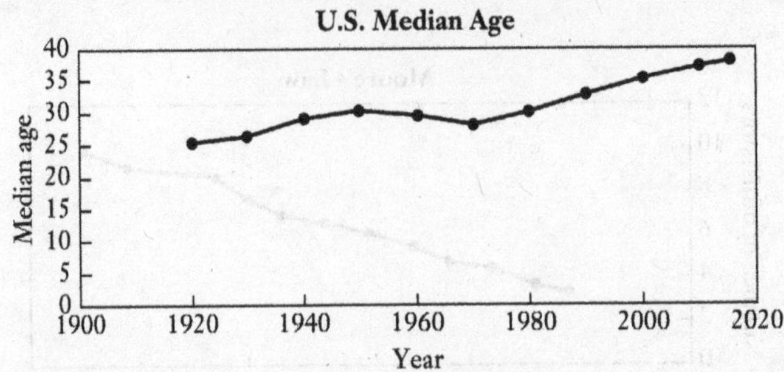

51. Answers will vary.

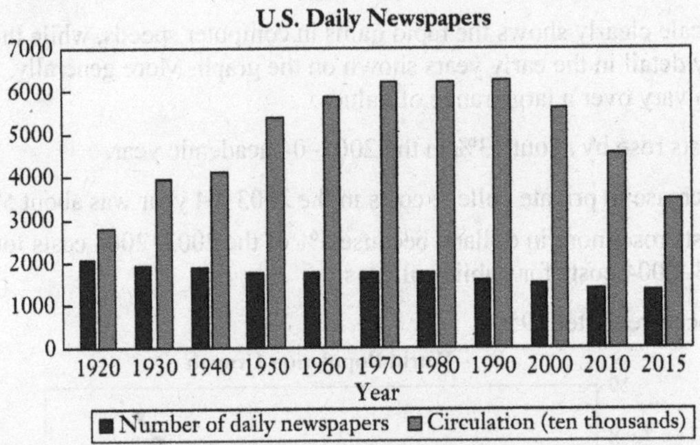

53. a. Answers will vary.

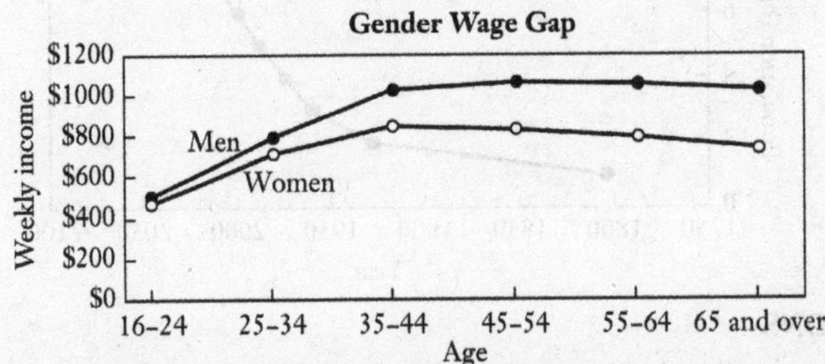

b. Wage ratios: ages 16–24, $\frac{\$470}{\$505} = 0.93$; ages 25–34, $\frac{\$710}{\$791} = 0.90$; ages 35–44, $\frac{\$845}{\$1024} = 0.83$; ages 45–54, $\frac{\$829}{\$1063} = 0.78$; ages 55–64, $\frac{\$795}{\$1054} = 0.75$; age 65+, $\frac{\$736}{\$1032} = 0.71$; The ratio decreases as age increases.

TECHNOLOGY EXERCISES

61.

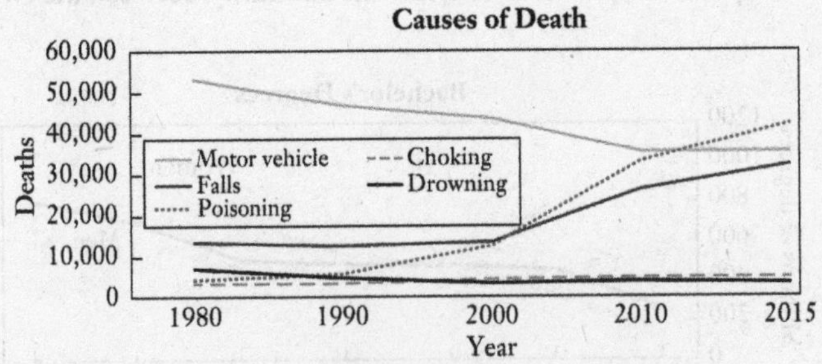

63.

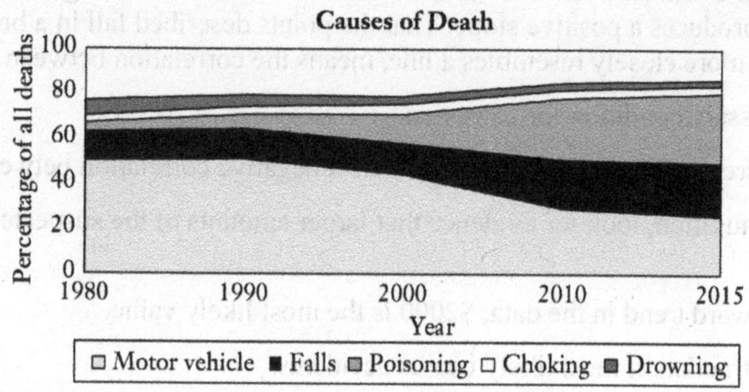

65.

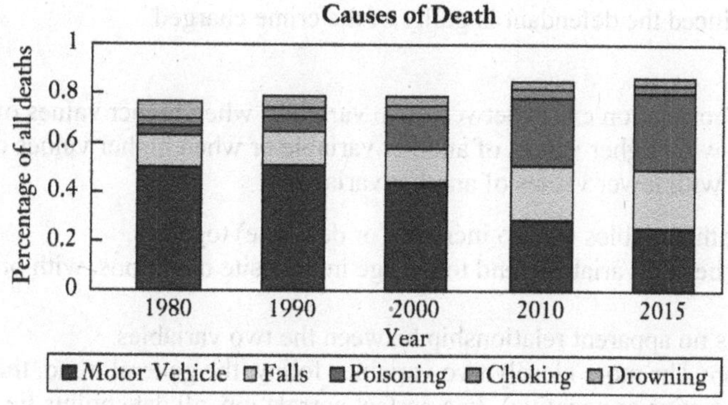

67. a.

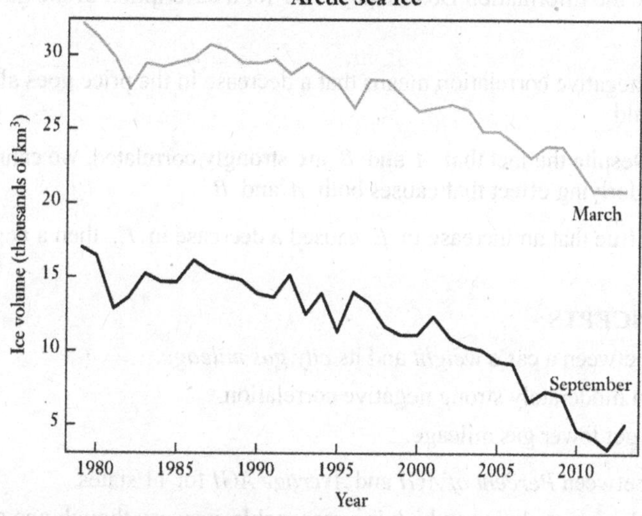

b. Answers will vary. Some observations might include that sea ice volume has been decreasing and that the graphs for the two months have similar shapes.

UNIT 5E: CORRELATION AND CAUSALITY

QUICK QUIZ

1. c. When one variable is correlated with another, an increase in one goes with either increasing values of the other, or decreasing values of the other.

2. b. The dot representing Brazil has a horizontal coordinate of about 62 years.

3. a. When points on a scatter diagram all lie near a line, there is a strong correlation between the variables represented. We call it a negative correlation when an increase in one variable goes along with a decrease in another (and this is what produces a negative slope).

4. **b**. A positive correlation exists between two variables when an increase in one goes with an increase in the other, and this is what produces a positive slope. That the points described fall in a broad swath, rather than a tight configuration that more closely resembles a line, means the correlation between the variables is weak.

5. **a**. Correlation is a necessary condition for causation.

6. **c**. Exercise tends to decrease body fat, so there would be a negative correlation between body fat and BMI.

7. **a**. When establishing causation, look for evidence that larger amounts of the suspected cause produce larger amounts of the effect.

8. **a**. Extrapolating the upward trend in the data, $2000 is the most likely value.

9. **c**. Since texting is a distraction, it most likely causes accidents.

10. **c**. The courts have proclaimed that a member of the jury should find the defendant guilty when the jury member is firmly convinced the defendant is guilty of the crime charged.

REVIEW QUESTIONS

1. Examples will vary. A correlation exists between two variables when higher values of one variable are consistently associated with higher values of another variable or when higher values of one variable are consistently associated with lower values of another variable.

3. Positive correlation: Both variables tend to increase (or decrease) together.
Negative correlation: The two variables tend to change in opposite directions, with one increasing while the other decreases.
No correlation: There is no apparent relationship between the two variables.
Strength of a correlation: The more closely two variables follow the general trend, the stronger the correlation (which may be either positive or negative). In a perfect correlation, all data points lie on a straight line.

5. Examples will vary. See the Information Box on page 361 for a description of the guidelines.

DOES IT MAKE SENSE?

7. Makes sense. A strong negative correlation means that a decrease in the price goes along with an increase in the number of tickets sold.

9. Does not make sense. Despite the fact that A and B are strongly correlated, we cannot be sure that A causes B (there may be an underlying effect that causes both A and B

11. Makes sense. If it were true that an increase in E caused a decrease in F, then a negative correlation would exist.

BASIC SKILLS AND CONCEPTS

13. We seek a correlation between a car's *weight* and its *city gas mileage*.
    a. The diagram shows a moderately strong negative correlation.
    b. Heavier cars tend to get lower gas mileage.

15. We seek a correlation between *Percent of AGI* and *Average AGI* for 11 states.
    a. At first glance, there's no correlation (which is a reasonable answer), though one could also argue for a very weak positive correlation based on the slight increasing trend of the data points.
    b. Higher AGI may imply slightly higher charitable giving as a percentage of AGI.

17. Use degrees of latitude and degrees of temperature as units. There is a strong negative correlation, because as you move north from the equator, the temperatures in June tend to decrease.

19. Use years and hours as units. There is probably a negative correlation because older people most likely do not spend as much time on social networking sites.

21. Use years and square miles as units. There is probably a positive correlation because states that joined the union later are usually larger than those that joined earlier. (Compare Alaska to Rhode Island, for example.)

23. Use children per woman and years as units. There is a strong negative correlation, because an increase in the life expectancy of a country goes along with a decrease in the number of children born.

FURTHER APPLICATIONS

25. a.

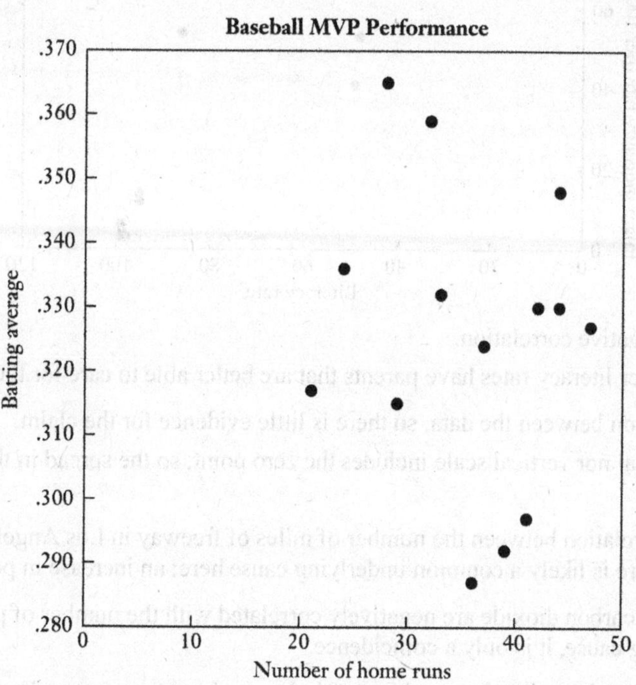

b. There does not appear to be any correlation.

c. Batting indicates the ability to get on base, not the ability to hit home runs.

27. a.

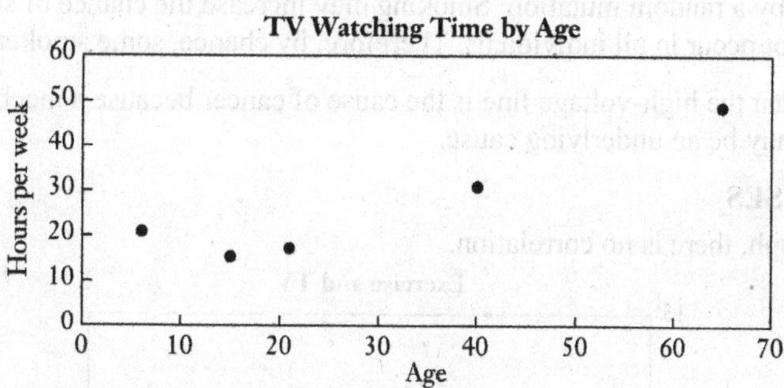

b. There is a strong positive correlation between the variables.

c. Older people are more likely to watch more traditional forms of media.

29. a.

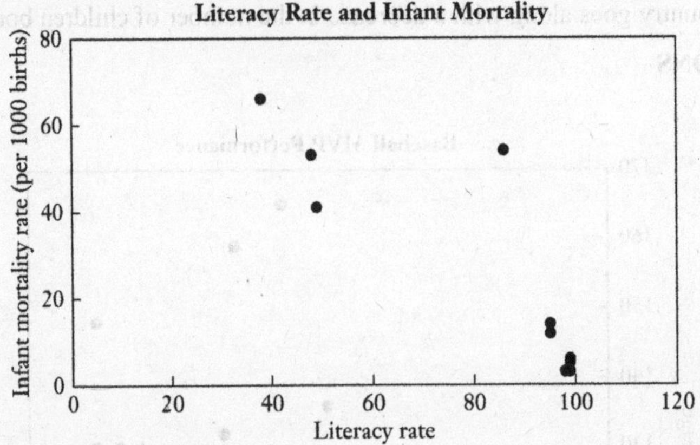

b. There is a strong negative correlation.

c. Countries with higher literacy rates have parents that are better able to care for their children.

31. a. There is no correlation between the data, so there is little evidence for the claim.

b. Neither the horizontal nor vertical scale includes the zero point, so the spread in the data is exaggerated for both variables.

33. There is a positive correlation between the number of miles of freeway in Los Angeles and the amount of traffic congestion. There is likely a common underlying cause here: an increase in population.

35. Levels of atmospheric carbon dioxide are negatively correlated with the number of pirates worldwide. There is no common underlying cause, it is only a coincidence.

37. The number of bartenders is positively correlated with the number of ministers. There may be a common underlying cause: an increase in the population of the city.

39. The causes of cancer are often random. A cancer cell is produced when the growth control mechanisms of a normal cell are altered by a random mutation. Smoking may increase the chance of such a mutation occurring, but the mutation will not occur in all individuals. Therefore, by chance, some smokers escape cancer.

41. You cannot conclude that the high-voltage line is the cause of cancer because a mechanism for the cause must be established. There may be an underlying cause.

TECHNOLOGY EXERCISES

49. a. According to the graph, there is no correlation.

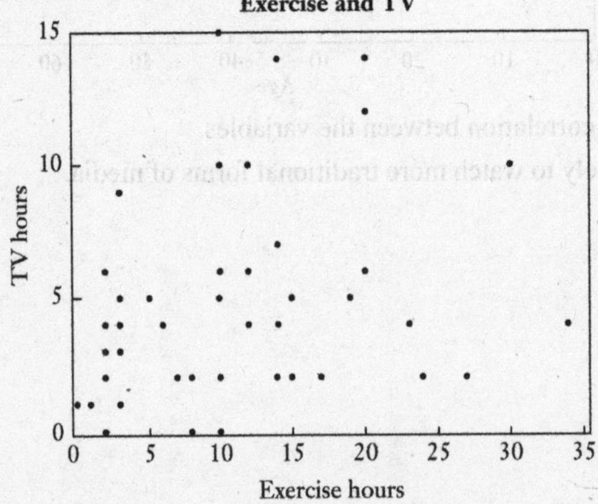

b. The correlation coefficient is 0.282.

UNIT 6A: CHARACTERIZING DATA

QUICK QUIZ

1. **a.** The median of a data set is the middle value when the data are placed in numerical order.
2. **b.** The median of a data set is the middle value when the data are placed in numerical order and the score of 79 is in the lower half of the data.
3. **c.** Outliers are data values that are much smaller or larger than the other values.
4. **a.** Outliers that are larger than the rest of the data set tend to pull the mean to the right of the median, and this is the case for all possible data sets that satisfy the conditions given.
5. **c.** Because the mean is significantly higher than the median, the distribution is skewed to the right, which implies some students drink considerably more than 12 sodas per week. Specifically, it can be shown that at least one student must drink more than 16 sodas per week.
6. **c.** The high salaries of a few superstars pull the tail of this distribution to the right.
7. **a.** High outliers tend to pull the mean to the right of the median, and thus you would hope to get a salary near the mean. There are data sets with high outliers that can be constructed where the median is higher than the mean, but these are rare.
8. **a.** In general, data sets with a narrow central peak typically have less variation than those with broad, spread out values (the presence of outliers can confound the issue).
9. **b.** The elite female gymnasts would be more uniform their weights than the randomly selected women.
10. **b.** The mayor would like to see a high median and low variation because that means there's a tight, central peak in the distribution that corresponds to support for her.

REVIEW QUESTIONS

1. The mean is what we most commonly call the average value. It is found using $\text{mean} = \dfrac{\text{sum of all values}}{\text{total number of values}}$. It can be considered the balance point of the distribution and is used with symmetric data. If the number of values is odd, the median is the middle value in the sorted data set (in either ascending or descending order), or halfway between the two middle values if the number of values is even. It is usually used with skewed data. The mode is the most common value (or group of values) in a distribution.

3. The different meanings of "average" can lead to confusion. Sometimes this confusion arises because we are not told whether the "average" is the mean or the median, and other times because we are not given enough information about how the average was computed.

5. Examples will vary. A single-peaked distribution is symmetric if its left half is a mirror image of its right half. A single-peaked distribution is left-skewed if its values are more spread out on the left side of the mode. A single-peaked distribution is right-skewed if its values are more spread out on the right side of the mode.

DOES IT MAKE SENSE?

7. Makes sense. The two highest grades may be large enough to balance the remaining seven lower grades so that the third highest is the mean.

9. Makes sense. When very high outliers are present, they tend to pull the mean to the right of the median (right-skewed).

11. Does not make sense. When the mean, median, and mode are all the same, it's a sign that you have a symmetric distribution of data.

BASIC SKILLS AND CONCEPTS

13. The mean, median, and mode are 354.6 words, 355.0 words, and 345 words, respectively.

15. The mean, median and mode are 0.188, 0.165, and 0.16, respectively. Because there is an even number of data points, the median is calculated by taking the mean of the middle two values (add 0.16 and 0.17, and divide by 2). The mode is 0.16.

17. The mean and median are 58.3 sec and 55.5 sec, respectively. Because there is an even number of data points, the median is calculated by taking the mean of the middle two values (add 53 and 58, and divide by 2). The mode is 49 sec.

19. The mean and median are 0.81237 and 0.8161, respectively. The outlier is 0.7901, because it varies from the others by a couple hundredths of a pound, while the rest vary from each other only by a few thousandths of a pound. Without the outlier, the mean is 0.81608, and the median is 0.8163.

21. The median would be a better representation of the average. A small number of people with high earnings would skew the distribution to the right, affecting the mean.

23. This distribution is probably skewed to the right by a small percentage of people who change jobs many times. Thus the median would do a better job of representing the average, as the outliers affect the mean.

25. This is a symmetric distribution, so the mean would be the best value to represent the average.

27. a. There is one peak on the right due to all the students who received A's.

    b.

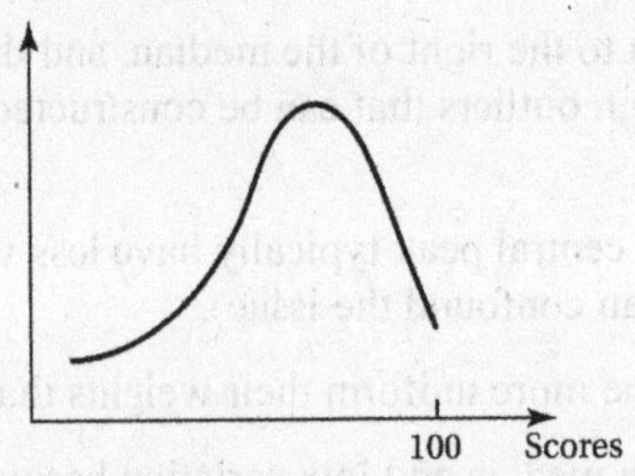

    c. The distribution is left-skewed because the scores are more spread out to the left.

    d. The variation is large.

29. a. There is likely one peak since most cities would have similar amounts of annual snowfall.

    b.

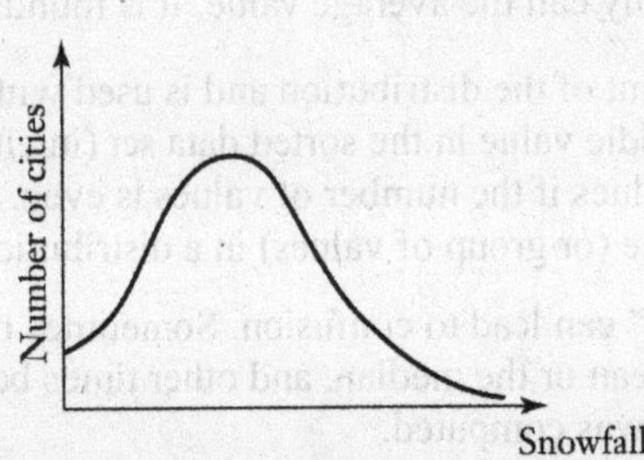

    c. The distribution would be right-skewed because a few cities with higher snowfall would vary from the average more than cities with lower snowfall.

    d. The variation would be large since some cities would have snowfall amounts much lower or much higher than the mean snowfall amount.

31. a. There would be one peak since sales would be high in the winter months, with very low sales (or even no sales) in the summer.

    b.

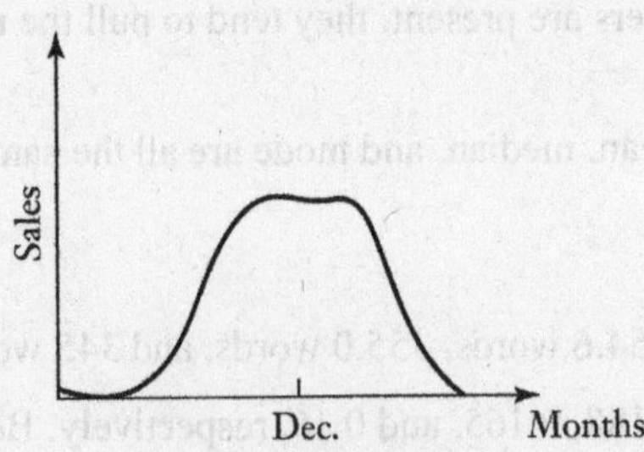

    c. This is a symmetric distribution.

    d. The variation is moderate because most of the sales occur in the winter.

33. a. There would be one peak in the distribution, corresponding to the average price of dog food.

b.

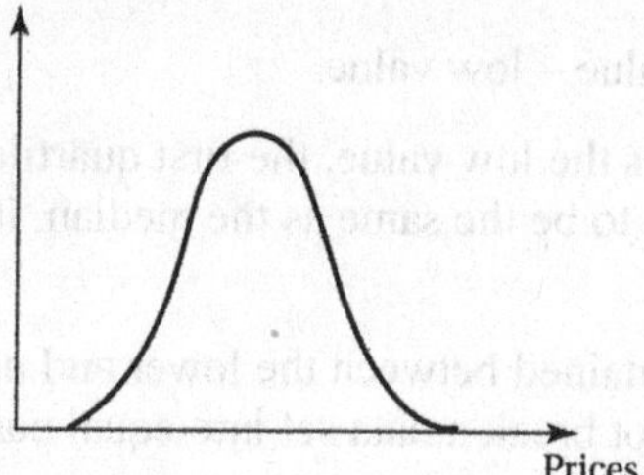

c. More likely than not, there would be no extreme outliers in this data set, and it would be symmetric.

d. There would be moderate variation in the data set because we do not expect outliers. (The dog food may not sell well next to 19 other brands if its price is radically different. Of course this begs the question, "But what if it's much cheaper than the other brands?" That's not too likely to happen for a commodity like dog food, because once one company begins selling its dog food at a low price, the others would be forced to lower their prices to compete).

35. The distribution would be symmetric since the mean is equal to the median. The variation would be somewhat high given the low score of 60 and the high score of 90, which are both far from the average score.

FURTHER APPLICATIONS

37. The distribution has two peaks (bimodal), with no symmetry, and large variation.

39. This distribution has one peak, is symmetric, and has moderate variation.

41. a. Sketches will vary. We cannot be sure of the exact shape of the distribution with the given parameters, as we do not have all the raw data. However, because the mean is larger than the median, and because the data set has large outliers, it would likely be right-skewed, with a single peak at the mode.

b. About 150 families (50%) earned less than $45,000 because that value is the median income.

c. We do not have enough information to be certain how many families earned more than $55,000 because we do not have the raw data, but it is likely a little less than half earned more than $55,000, simply because $55,000 is greater than the median.

43. a. $(4\times4.0+3\times3.4+4\times2.4+3\times2.7+1\times3.7)\div(4+3+4+3+1)=47.6\div15=3.17$

b. Let F be the required grade in French.

$$(4\times4.0+3\times3.4+4\times F+3\times2.7+1\times3.7)\div(4+3+4+3+1)=3.5$$
$$(4F+38.0)\div15=3.5$$
$$4F+38.0=52.5$$
$$4F=14.5$$
$$F=3.6$$

She would need at least an A– in French.

TECHNOLOGY EXERCISES

51. a. The mean, median and mode for team A are 134, 125, and 115 respectively. The mean and median for team B are 131 and 135, respectively. There is no mode for Team B. The coach of team A can claim that his team has the greater mean weight. The coach of team B can claim that his team has the greater median weight.

b. The mean weight of the teams combined is 132.5, which is equal to the mean of the two mean weights of the teams. This is true since both teams have the same number of players.

c. The median weight of the teams combined is 130, which is equal to the median of the two median weights of the teams. This result is not true in general.

53. Answers will vary.

UNIT 6B: MEASURES OF VARIATION

QUICK QUIZ

1. **a.** The range is defined as the high value – low value.

2. **c.** The five-number summary includes the low value, the first quartile, the median, the third quartile, and the high value. Unless the mean happens to be the same as the median, it would not be part of the five-number summary.

3. **a.** Roughly half of any data set is contained between the lower and upper quartiles ("roughly" because with an odd number of data values, one cannot break a data set into equal parts).

4. **b.** Consider the set {0, 100, 100, 100, 100, 100, 100, 100, 100, 100}. Its mean is 90, which is smaller than the lower quartile of 100.

5. **b.** You need the high and low values to compute the range, and you need all of the values to compute the standard deviation, so the only thing you can compute is a single deviation.

6. **b.** The standard deviation is defined in such a way that it can be interpreted as the average distance of a random value from the mean.

7. **c.** The standard deviation is defined as the square root of the variance, and thus is always non-negative.

8. **a.** Using the range rule of thumb, we see that the range is about four times as large as the standard deviation. (The worst-case scenario is when the range and standard deviation are both zero).

9. **b.** Newborn infants and first grade boys have similar heights, whereas there is considerable variation in the heights of all elementary school children.

10. **c.** Because the standard deviation for Garcia is the largest, the data are more spread out, so Professor Garcia must have had very high grades from more students than the other two.

REVIEW QUESTIONS

1. The waiting times at the grocery store with the larger variation would vary over a fairly wide range, so a few customers could have long waits and are likely to become annoyed. In contrast, customers at the grocery store with the smaller variation would probably feel they are waiting roughly equal times.

3. The lower quartile (or first quartile) divides the lowest fourth of a data set from the upper three-fourths. It is the median of the data values in the lower half of a data set. (Exclude the middle value in the data set if the number of data points is odd.)
   The middle quartile (or second quartile) is the median of the entire data set.
   The upper quartile (or third quartile) divides the lowest three-fourths of a data set from the upper fourth. It is the median of the data values in the upper half of a data set. (Exclude the middle value in the data set if the number of data points is odd.)

5. Examples will vary. See the Information Box on page 390. If the sample values are all the same, the standard deviation is zero.

DOES IT MAKE SENSE?

7. Does not make sense. The range depends upon only the low and high values, not on the middle values.

9. Makes sense. Consider the case where the 15 highest scores were 80, the next score was 68, and the lowest 14 scores were 40. There are numerous such cases that satisfy the conditions given.

11. Makes sense. The standard deviation describes the spread of the data, and one would certainly expect the heights of 5-year old children to have less variation than the heights of children aged 3 to 15.

BASIC SKILLS AND CONCEPTS

13. The mean waiting time at Big Bank is $(4.1+5.2+5.6+6.2+6.7+7.2+7.7+7.7+8.5+9.3+11.0)/11$
    $= 79.2/11 = 7.2$. Because there are 11 values, the median is the sixth value, which is 7.2.

15. a. For the high-cost cities, the mean, median, and range are 166.0, 149, and 83, respectively. The mean, median, and range for the low-cost cities are 80.7, 81, and 7, respectively.

b. For the high-cost cities, the five-number summary is (145, 148, 149, 177, 228). For the low-cost cities, it is (76, 80, 81, 83, 83).

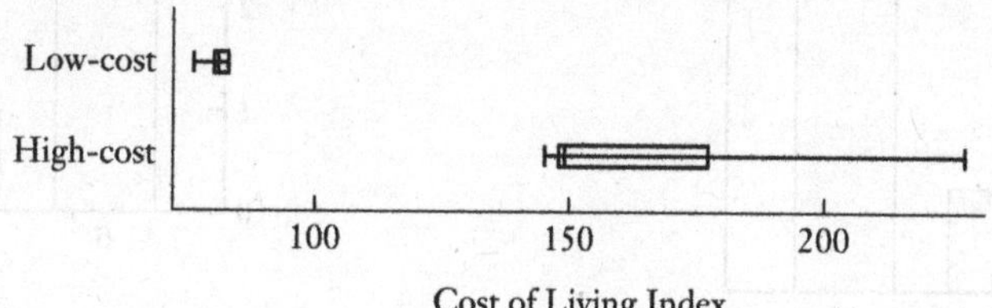

c. The standard deviation for the high-cost cities is 32.6. It is 2.7 for the low-cost cities.

d. The range rule of thumb estimates the standard deviation for the high-cost cities as $83 \div 4 = 20.8$, which underestimates the actual value of 32.6. For the low-cost cities, the range rule of thumb estimates the standard deviation as $7 \div 4 = 1.8$, which also underestimates the actual value of 2.7.

e. The mean and median cost of living index for low-cost cities are considerably smaller than the cost of living index in high-cost cities, though the variation is much smaller.

17. a. For the National League (NL), the mean, median, and range are 0.5217, 0.5105, and 0.220, respectively. For the American League (AL), the mean, median, and range are 0.5005, 0.5245, and 0.222, respectively.

b. The five-number summary for the National League is (0.420, 0.463, 0.5105, 0.586, 0.640). It is (0.364, 0.420, 0.5245, 0.584, 0.586) for the American League.

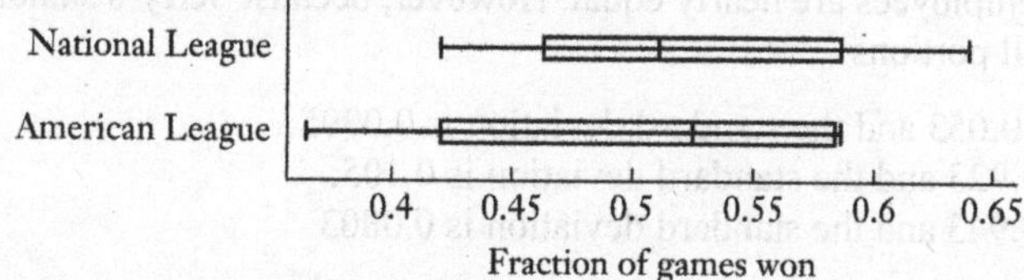

c. The standard deviation for the National League is 0.0819; for the American League, it is 0.0914.

d. The range rule of thumb estimates the standard deviation for the National League as $0.220 \div 4 = 0.0550$, which underestimates the actual value of 0.0819. For the American League, the range rule of thumb estimates the standard deviation as $0.222 \div 4 = 0.0555$, which is underestimates the actual value of 0.0914.

e. The average weight and standard deviation of fractions of games won are roughly the same for the National League and the American League.

FURTHER APPLICATIONS

19. a.

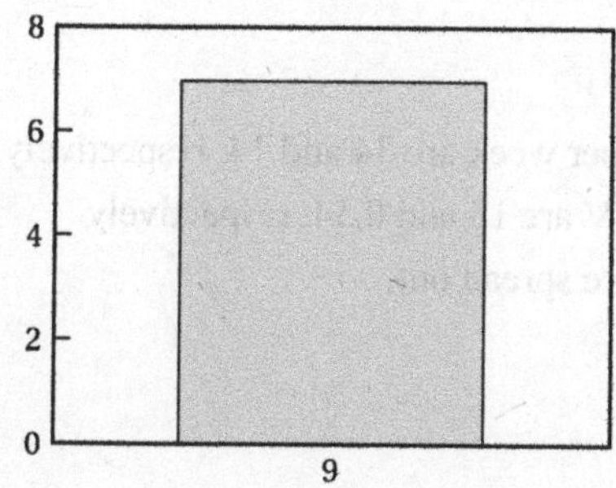

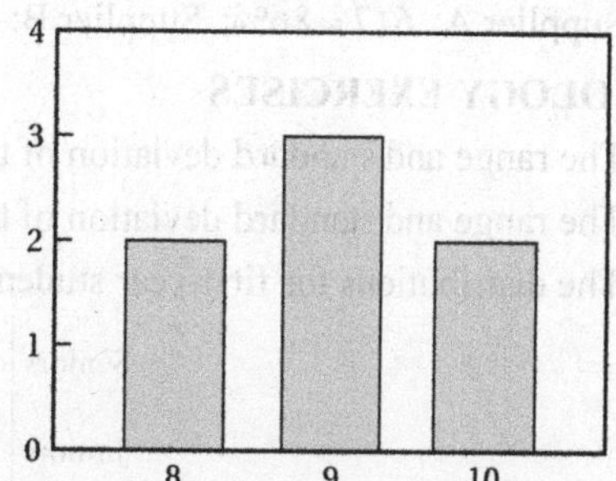

19. (continued)

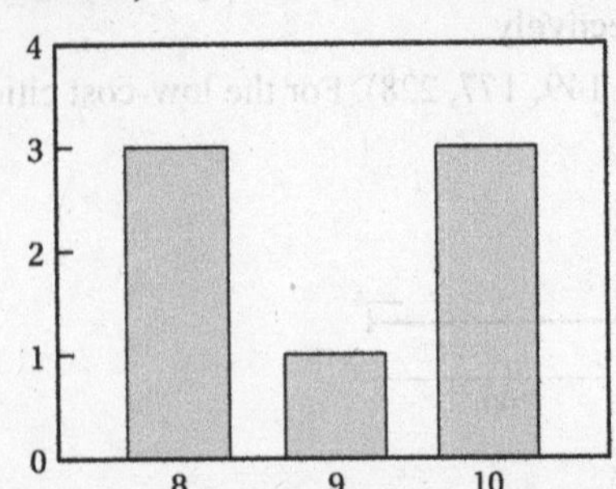

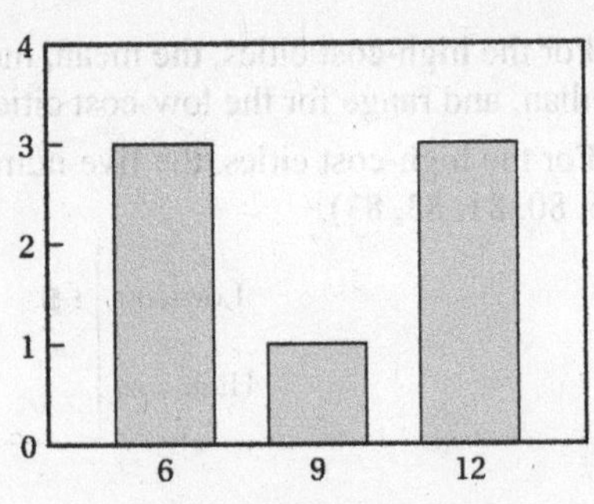

b. The five-number summaries for each of the sets shown are (in order): (9, 9, 9, 9, 9); (8, 8, 9, 10, 10); (8, 8, 9, 10, 10); and (6, 6, 9, 12, 12). The boxplots are not shown.

c. The standard deviations for the sets are (in order): 0.000, 0.816, 1.000, and 3.000.

d. Looking at just the answers to part c, we can tell that the variation gets larger from set to set.

21. The first pizza shop has a slightly larger mean delivery time, but much less variation in delivery time when compared to the second shop. If you want a reliable delivery time, choose the first shop. If you do not care when the pizza arrives, choose the shop that offers cheaper pizza (you like them equally well, so you may as well save some money).

23. A lower standard deviation means more certainty in the return, and a lower risk.

25. The means of the two employees are nearly equal. However, because Jerry's standard deviation is larger, he likely serves more small portions.

27. a. 2000: The mean is 10.053 and the standard deviation is 0.0995.
2008: The mean is 9.923 and the standard deviation is 0.105.
2016: The mean is 9.943 and the standard deviation is 0.0803

b. Answers will vary

29. a. For Supplier A, the mean and standard deviation are 16.32 mm and 0.057 mm, respectively. For Supplier B, the mean and standard deviation are 16.33 mm and 0.098 mm, respectively.

b.

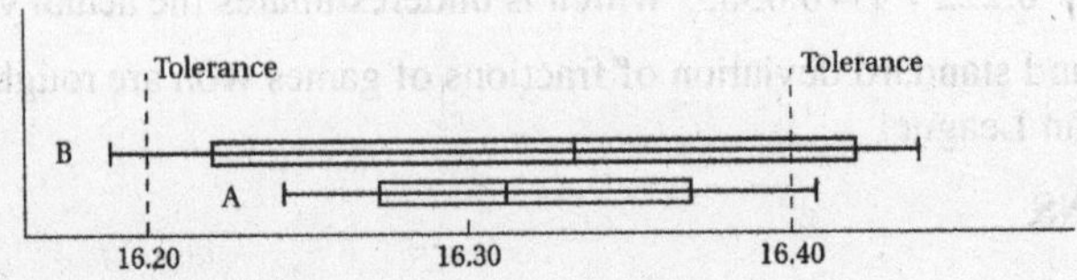

c. Supplier A: $6/7 \approx 86\%$; Supplier B: $4/7 \approx 57\%$

TECHNOLOGY EXERCISES

35. a. The range and standard deviation of the hours spent exercising per week are 34 and 14, respectively.

b. The range and standard deviation of the hours spent watching TV are 15 and 0.54, respectively.

c. The distributions for first-year students and sophomores are more spread out.

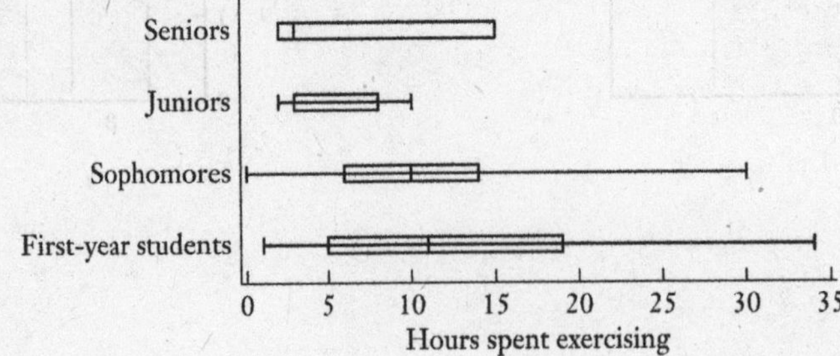

35. (continued)

d. The distributions are similar.

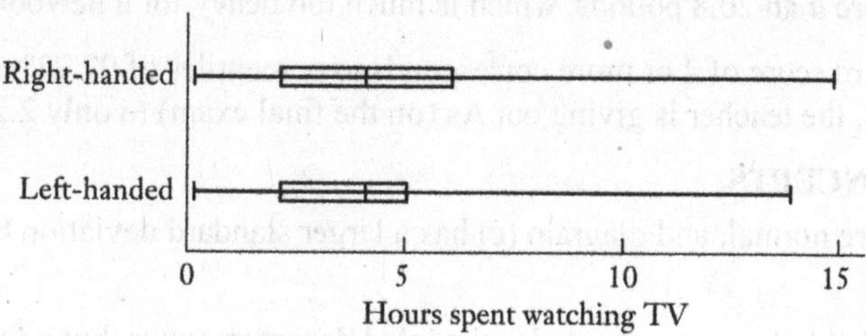

UNIT 6C: THE NORMAL DISTRIBUTION

QUICK QUIZ

1. b. A normal distribution is symmetric with one peak, and this produces a bell-shaped curve.
2. a. The mean, median and mode are equal in a normal distribution.
3. a. Data values farther from the mean correspond to lower frequencies than those close to the mean, due to the bell-shaped distribution.
4. c. Since most of the workers earn minimum wage, the mode of the wage distribution is on the left, and the distribution is right-skewed.
5. a. Roughly 68% of data values fall within one standard deviation of the mean, and 68% is about 2/3.
6. c. In a normally distributed data set, 99.7% of all data fall within 3 standard deviations of the mean.
7. a. Note that 43 mpg is 1 standard deviation below the mean. Since 68% of the data lies within one standard deviation of the mean, the remaining 32% is farther than 1 standard deviation from the mean. The normal distribution is symmetric, so half of 32%, or 16%, lies below 1 standard deviation.
8. c. The z-score for 84 is $\dfrac{84-75}{6}=1.5$.
9. c. If your friend said his IQ was in the 75th percentile, it would mean his IQ is larger than 75% of the population. It is impossible to have an IQ that is larger than 102% of the population.
10. c. Table 6.4 shows that the percentile corresponding with a standard score of -0.50 is 30.15%, which means 30.15% of girls measured were the same height or shorter than this particular girl.

REVIEW QUESTIONS

1. Examples will vary. The normal distribution is a symmetric, bell-shaped distribution with a single peak. Its peak corresponds to the mean, median, and mode of the distribution. Its variation is characterized by the standard deviation of the distribution. A data set is likely to have a nearly normal distribution if most data values are clustered near the mean, the data values are spread evenly around the mean, larger deviations from the mean become increasingly rare, and individual data values result from a combination of many different factors, such as genetic and environmental factors.

3. The standard score (*z-score*) gives the number of standard deviations a data value lies above or below the mean. The standard score is found using $z = \text{standard score} = \dfrac{\text{data value} - \text{mean}}{\text{standard deviation}}$. The standard score is positive for data values above the mean and negative for data values below the mean.

DOES IT MAKE SENSE?

5. Makes sense. Physical characteristics are often normally distributed, and college basketball players are tall, though not all of them, so it makes sense that the mean is 6'8", with a standard deviation of 4" (this would put approximately 95% of the players within 6'0" to 7'6").

7. Does not make sense. With a mean of 6.8 pounds, and a standard deviation of 7 pounds, you would expect that a small percentage of babies would be born 2 standard deviations above the mean. But this would imply that some babies weigh more than 20.8 pounds, which is much too heavy for a newborn.

9. Makes sense. A standard score of 2 or more corresponds to percentiles of 97.72% and higher, so while this could certainly happen, the teacher is giving out As (on the final exam) to only 2.28% of the class

BASIC SKILLS AND CONCEPTS

11. Diagrams (a) and (c) are normal, and diagram (c) has a larger standard deviation because its values are more spread out.

13. Most trains would probably leave near to their scheduled departure times, but a few could be considerably delayed, which would result in a right-skewed (and non-normal) distribution.

15. Random factors would be responsible for the variation seen in the weights of the pickup trucks, so this distribution would be normal.

17. Easy exams would have scores that are left-skewed, so the scores would not be normally distributed.

19. a. Half of any normally distributed data set lies above the mean, so 50% of the scores are greater than 100.

    b. Note that 120 is one standard deviation above the mean. Since 68% of the data lies within one standard deviation above the mean (between 80 and 120), half of 68%, or 34%, lies between 100 and 120. Subtract 34% from 50% (see part a) to find that 16% of the scores are above 120.

    c. A sketch of the normal distribution would show that the amount of data lying above 120 is equivalent to the amount of data lying below 80, so the answer is 16% (see part b).

    d. Note that 140 is two standard deviations above the mean. Since 95% of the data lies within two standard deviations above the mean (between 60 and 140), half of 95%, or 47.5%, lies between 100 and 140. Add 47.5% to 50% (see part a) to find that 97.5% of the scores are below 140.

    e. Note that 60 is two standard deviations below the mean. Since 95% of the data lies within two standard deviations of the mean (between 60 and 140), 5% of the data lies outside this range. Because the normal distribution is symmetric, half of 5%, or 2.5%, lies in the lower tail below 60.

    f. Note that 120 is one standard deviation above the mean. Since 68% of the data lies within one standard deviation of the mean (between 80 and 120), half of 68%, or 34%, lies between 100 and 120. Add 34% to 50% (the answer to part a) to find that 84% of the scores are less than 120.

    g. A sketch of the normal distribution shows that the percent of data lying above 80 is equivalent to the percent lying below 120, so the answer is 84% (see part b).

    h. The percent of data between 80 and 120 is 68%, because these are each one standard deviation from the mean.

21. A score of 59 is one standard deviation below the mean. Since 68% of normally distributed data lies within one standard deviation of the mean, half of 68%, or 34%, lies between 59 and 67 due to symmetry. The percent of data lying below 67 is 50%, so we subtract these two results to find that $50\% - 34\% = 16\%$ of the scores lie below 59.

23. Subtract two standard deviations from the mean to get the cutoff score of $67 - 16 = 51$. Since 95% of the data lies within two standard deviations of the mean, 5% lies outside this range. Half of 5%, or 2.5%, will be below 51, and will fail the exam.

25. The standard score for 67 is $z = \frac{67-67}{8} = 0$.

27. The standard score for 79 is $z = \frac{79-67}{8} = 1.5$.

29. a. $z = -1$; The corresponding percentile is 15.87%.

    b. $z = -0.5$; The corresponding percentile is 30.85%.

    c. $z = 1.5$; The corresponding percentile is 93.32%.

31. a. The 30th percentile corresponds to a standard score of about $z = -0.52$, which means the data value is about 0.52 standard deviations below the mean.

b. The 70th percentile corresponds to a standard score of about $z = 0.52$, which means the data value is 0.52 standard deviations above the mean.

c. The 55th percentile corresponds to a standard score of about $z = 0.12$, which means the data value is 0.12 standard deviations above the mean.

FURTHER APPLICATIONS

33. About 68% of births occur within 15 days of the mean, because this range is one standard deviation from the mean.

35. Since 68% of births occur within 15 days of the due date, 32% occur outside this range, and half of 32%, or 16%, occur more than 15 days after the due date.

37. a. The standard score is $z = \frac{64-63.6}{2.5} = 0.16$, which corresponds to an approximate percentile of 56.4%.

b. $z = \frac{61-63.6}{2.5} = -1.04$, which corresponds to an approximate percentile of 14.9%.

c. $z = \frac{59.8-63.6}{2.5} = -1.52$, which corresponds to an approximate percentile of 6.4%.

d. $z = \frac{64.8-63.6}{2.5} = 0.48$, which corresponds to an approximate percentile of 68.4%.

39. It is not likely because heights of eighth-graders would be normally distributed, and a standard deviation of 40 inches would imply some of the eighth graders (about 16%) are taller than 95 inches, which is preposterous.

41. The 90th percentile corresponds to a *z*-score of about 1.3, which means a data value is 1.3 standard deviations above the mean. For the verbal portion of the GRE, this translates into a score of $150 + 1.3(8.5) = 161$. For the quantitative portion, it translates into a score of $152 + 1.3(8.9) = 164$.

43. The standard score for 160 is $z = \frac{160-152}{8.9} = 0.90$, which corresponds to about the 81.6th percentile. This is the percent of students who score below 160, so the percent that score above 160 is 18.4%.

45. a. The standard score for 150 is $z = \frac{150-152}{8.9} = -0.225$, which corresponds to about the 41.1 percentile.

b. The standard score for 157 is $z = \frac{157-150}{8.5} = 0.824$, which corresponds to about the 79.5 percentile.

47. The standard score for 130 is $z = \frac{130-150}{8.5} = -2.353$, which corresponds to about the 0.9 percentile. The standard score for 170 is $z = \frac{170-150}{8.5} = 2.353$, which corresponds to about the 99.1 percentile.

TECHNOLOGY EXERCISES

51. Answers will vary.

UNIT 6D: STATISTICAL INFERENCE

QUICK QUIZ

1. **c.** The relative frequencies of 38 or fewer heads are extremely low.

2. **b.** In order to determine whether a result is statistically significant, we must be able to compare it to a known quantity (such as the effect on a control group in this case).

3. **a.** Statistical significance at the 0.01 level means an observed difference between the effectiveness of the remedy and the placebo would occur by chance in fewer than 1 out of 100 trials of the experiment.

4. **c.** If the statistical significance is at the 0.05 level, we would expect that 1 in 20 experiments would show the pill working better than the placebo by chance alone.

5. **b.** The confidence interval is found by subtracting and adding 3% to the results of the poll (35%).

6. **c.** The central limit theorem tells us that the true proportion is within the margin of error 95% of the time.

7. **c.** The margin of error is approximately $1/\sqrt{n}$, Set this equal to 4% (0.04), and solve for n to find that the sample size was 625. If we quadruple the sample size to 2500, the margin of error will be approximately $1/\sqrt{2500} = 0.02$, or 2%.

8. **a.** The null hypothesis is often taken to be the assumption that there is no difference in the things being compared.

9. **c.** If you cannot reject the null hypothesis, the only thing you can be sure of is that you don't have evidence to support the alternative hypothesis.

10. **b.** If the difference in gas prices can be explained by chance in only 1 out of 100 experiments, you've found a result that is statistically significant at the 0.01 level.

REVIEW QUESTIONS

1. Statistical inference uses results from a sample to *infer* a conclusion about the population that the sample was obtained from. It allows us to make conclusions without having to exam an entire population.

3. If the results from a sample are not statistically significant, they could have occurred by random chance. If this is the case, the results of the sample cannot be used to make a claim about the population.

5. Examples will vary. The margin of error for a 95% confidence interval is given by $1/\sqrt{n}$. For a given sample, the lower limit of the 95% confidence interval is found by subtracting the margin of error from the sample proportion. The upper limit is found by adding the margin of error to the sample proportion.

7. The two possible outcomes of a hypothesis test are either reject the null hypothesis or do not reject the null hypothesis. Accepting the null hypothesis is not a possible outcome, because the null hypothesis is the starting assumption.

DOES IT MAKE SENSE?

9. Makes sense. If the number of people cured by the new drug is not significantly larger than the number cured by the old drug, the difference could be explained by chance alone, which means the results are not statistically significant.

11. Does not make sense. The margin of error is based upon the number of people surveyed (the sample size), and thus both Agency A and B should have the same margin of error.

13. Makes sense. The alternative hypothesis is the claim that is accepted if the null hypothesis is rejected.

BASIC SKILLS AND CONCEPTS

15. Most of the time you would expect around 20 sixes in 60 rolls of a standard die. It would be very rare to get only 2 sixes, and thus this is statistically significant.

17. Note that 19/400 is about 5%, which is exactly what you would expect by chance alone from an airline that has 95% of its flights on time. Thus, this is not statistically significant.

19. There are not many winners in the Power Ball lottery, and yet there are a good number of 7-11 stores. It would be very unlikely that ten winners in a row purchased their tickets at the same store, so this is statistically significant.

21. The result is significant at far less than the 0.01 level (which means it is significant at both the 0.01 and 0.05 level), because the probability of finding the stated results is less than 1 in 1 million by chance alone. This gives us good evidence to support the alternative hypothesis that the accepted value for temperature is wrong.

23. The results of the study are significant at neither the 0.05 level nor the 0.01 level, and thus the improvement is not statistically significant.

25. The margin of error is $1/\sqrt{1012} = 0.031$, or 3.1%, and the 95% confidence interval is $32.0\% - 3.1\% = 28.9\%$ to $32.0\% + 3.1\% = 35.1\%$. This means we can be 95% confident that the actual percentage is between 28.9% and 35.1%.

27. The margin of error is $1/\sqrt{2010} = 0.022$, or 2.2%, and the 95% confidence interval is $83.0\% - 2.2\% = 80.8\%$ to $83.0\% + 2.2\% = 85.2\%$. This means we can be 95% confident that the actual percentage is between 80.8% and 85.2%.

29. The margin of error is $1/\sqrt{1000} = 0.032$, or 3.2%, and the 95% confidence interval is $53.0\% - 3.2\% = 49.8\%$ to $53.0\% + 3.2\% = 56.2\%$. This means we can be 95% confident that the actual percentage is between and 49.8% and 56.2%.

31. The margin of error is $1/\sqrt{1549} = 0.025$, or 2.5%, and the 95% confidence interval is $82.0\% - 2.5\% = 79.5\%$ to $82.0\% + 2.5\% = 84.5\%$. This means we can be 95% confident that the actual percentage is between 79.5% and 84.5%.

33. a. Null hypothesis: six-year graduation rate = 60%. Alternative hypothesis: six-year graduation rate > 60%.

    b. Rejecting the null hypothesis means there is evidence that the graduation rate exceeds 60%. Failing to reject the null hypothesis means there is insufficient evidence to conclude that the graduation rate exceeds 60%.

35. a. Null hypothesis: mean teacher salary = $57,900. Alternative hypothesis: mean teacher salary > $57,900.

    b. Rejecting the null hypothesis means there is evidence that the mean teacher salary in the state exceeds $57,900. Failing to reject the null hypothesis means there is insufficient evidence to conclude that the mean teacher salary exceeds $57,900.

37. a. Null hypothesis: percentage of delayed flights = 15%. Alternative hypothesis: percentage of delayed flights > 15%

    b. Rejecting the null hypothesis means there is evidence that the percentage of delayed flights exceeds 15%. Failing to reject the null hypothesis means there is insufficient evidence to conclude that the percentage of delayed flights exceeds 15%.

39. Null hypothesis: mean annual mileage of cars in the fleet = 11,725 miles. Alternative hypothesis: mean annual mileage of cars in the fleet > 11,725. The result is significant at the 0.01 level, and provides good evidence for rejecting the null hypothesis.

41. Null hypothesis: mean stay = 2.1 days. Alternative hypothesis: mean stay > 2.1 days. The result is not significant at the 0.05 level, and there are no grounds for rejecting the null hypothesis.

43. Null hypothesis: mean income = $50,000. Alternative hypothesis: mean income > $50,000. The result is significant at the 0.01 level, and provides good evidence for rejecting the null hypothesis.

FURTHER APPLICATIONS

45. The margin of error is $1/\sqrt{5000} = 0.014$, or 1.4%, and the 95% confidence interval is $12.8\% - 1.4\% = 11.4\%$ to $12.8\% + 1.4\% = 14.2\%$. This means we can be 95% confident that the actual percentage of viewers who watched *NCIS* was between 11.4% and 14.2%.

47. The margin of error is approximately $1/\sqrt{n}$. If we want to decrease the margin of error by a factor of 2, we must increase the sample size by a factor of 4, because $\frac{1}{\sqrt{4n}} = \frac{1}{2\sqrt{n}} = \frac{1}{2} \times \frac{1}{\sqrt{n}}$.

49. The margin of error, 3%, is consistent with the sample size because $1/\sqrt{1019} = 0.03 = 3\%$.

TECHNOLOGY EXERCISES

55. a. Answers will vary.

    b. Answers will vary.

UNIT 7A: FUNDAMENTALS OF PROBABILITY

QUICK QUIZ

1. **b.** TTH is a different outcome than THT due to the order in which the heads appear, but both are the same event because each has two tails.
2. **b.** This is a probability based on observations about the number of free throws Lisa attempted and made, and so it is empirical.
3. **a.** This is a theoretical probability because we are assuming each outcome is equally likely, and no experiment is being carried out to determine the probability.
4. **c.** If the probability of an event is 0.001 the probability it does not occur is $1-0.001$.
5. **c.** The most likely outcome when a coin is tossed several times is that half of the time heads will appear.
6. **a.** Table 7.4 shows that the probability Serena wins is $5/36$, whereas the probability that Mackenzie wins is $4/36$.
7. **c.** There are six possible outcomes for each of the four dice, and the multiplication principle says we multiply these outcomes together to get the total number of outcomes when all four dice are rolled.
8. **c.** There are six possible outcomes for each of the three dice, so there are $6\times6\times6$ possible different outcomes.
9. **a.** The sum of all the probabilities for a probability distribution is always 1.
10. **a.** If Triple Treat's probability of winning is 1/5, then the probability he loses is 4/5. This implies the odds that he wins are $\frac{1/5}{4/5}=1/4$, or 1 to 4, which, in turn, means the odds against are 4 to 1.

REVIEW QUESTIONS

1. Outcomes are the most basic possible results of observations or experiments. For example, if you toss two coins, one possible outcome is HT and another possible outcome is TH. An event consists of one or more outcomes that share a property of interest. For example, if you toss two coins and count the number of heads, the outcomes HT and TH both represent the same event of 1 head (and 1 tail).

3. A theoretical probability is based on the assumption that all outcomes are equally likely. It is calculated by dividing the number of ways an event can occur by the total number of possible outcomes. A relative frequency probability is based on observations or experiments. It is the relative frequency of the event of interest. A subjective probability is an estimate based on experience or intuition.

5. A probability distribution represents the probabilities of all possible events of interest.
To make a table of a probability distribution:
First: List all possible outcomes.
Second: Identify outcomes that represent the same event. Find the probability of each event.
Third: Make a table or figure that displays all the probabilities. The sum of all the probabilities must be 1.

DOES IT MAKE SENSE?

7. Makes sense. The outcomes are HHTT, TTHH, THTH, THTH, THHT, THHT.

9. Makes sense. This is a subjective probability, but it's as good as any other guess one might make.

11. Does not make sense. The sum of $P(A)$ and $P(not\ A)$ is always 1.

BASIC SKILLS AND CONCEPTS

13. There are $12\times4=48$ choices.

15. There are $2\times8\times6=96$ three-course meals.

17. The possible outcomes for the two games are (WW, WL, LW, LL). If only concerned with the number of Wins, the outcomes are (0,1,2).

19. The possible outcomes for the two tosses are (HH, HT, TH, TT). The probability is 3/4. We are assuming equally likely outcomes.

21. There are 4 kings in a standard deck, so the probability is $4/52 = 1/13$. We are assuming that each card is equally likely to be chosen.

23. There are 4 possible outcomes with a 2-child family (BB, BG, GB, GG), and 1 of these have exactly two boys, so the probability is $1/4$. This assumes each outcome is equally likely to occur.

25. There are 12 months in the year, so the probability is $1/12$, assuming births are uniformly distributed across the months of the year. If using the number of days in the months, there are 30 total days in April, so the probability is $30/365 = 6/73$, assuming births are uniformly distributed across the days of the year.

27. There are 31 days in December, so the probability is $1/31$, assuming births are uniformly distributed across the days of the month.

29. The relative frequency probability is $15/27 = 5/9$.

31. A relative frequency probability of $15/100 = 0.15$ is not near the expected frequency probability of 0.5. The coin is likely not fair.

33. theoretical

35. The probability is $5/6$, because there are 5 outcomes that aren't 1s. It is assumed that each outcome is equally likely.

37. The probability he will make the next free throw is $69\% = 0.69$, so the probability he will miss is $1-0.69 = 0.31$, or 31%. The probabilities computed here are based on his past performance.

39.

Result	Probability
0 B	1/8
1 B	3/8
2 B	3/8
3 B	1/8

41.

Sum	Probability
2	1/16
3	2/16
4	3/16
5	4/16
6	3/16
7	2/16
8	1/16

43. The probability of getting a 5 is 1/6, so the probability of not getting a 5 is 5/6. Thus the odds for the event are $\frac{1/6}{5/6} = 1/5 =$ 1 to 5. The odds against are 5 to 1.

45. The probability of getting a double 6 is 1/36, so the probability of not getting a double 6 is 35/36. Thus the odds for the event are $\frac{1/36}{35/36} = 1/35 =$ 1 to 35. The odds against are 35 to 1.

47. Since the odds on your bet are 3 to 5, you'll win \$3 for every \$5 you bet. A \$20 bet is four \$5 bets, so you will gain \$12.

FURTHER APPLICATIONS

49. Theoretical: There are 6 red face cards, so the probability is $6/52 = 3/26$. It is assumed that each card is equally likely to be chosen.

51. Theoretical: The probability is $20/52 = 5/13$ because there are 20 even-numbered cards. We are assuming equally likely outcomes.

53. Assuming that people are equally likely to be born during any hour of the day, the probability is $2/24 = 1/12$.

55. Theoretical probability: There are ten possible outcomes for the last digit in a phone number, and two of these are 0 or 1, so the probability is $2/10 = 1/5$. It is assumed that each digit is equally likely to occur at the end of a phone number.

57. Theoretical probability: There are 8 possible outcomes with a 3-child family (BBB, BBG, BGB, GBB, GGB, GBG, BGG, GGG), and 1 of these has exactly three girls, so the probability is $1/8$. This assumes each outcome is equally likely to occur.

59. Relative frequency: There are ten possible outcomes for the last digit in a social security number, so the probability is $1/10$, assuming that final digits are uniformly distributed over social security numbers.

61. Relative frequency: The probability is $250/1000 = 0.25$.

63. Relative frequency: There are 15 odd days in June, so the probability is $1/15$. Equally likely outcomes are assumed.

65. Relative frequency: The probability of hearing your favorite group is $127/889 = 1/7$, so the probability of not hearing your favorite group is $1 - 1/7 = 6/7$, Equally likely outcomes are assumed.

67. a. {RR, RW, RB, WW, WR, WB, BB, BR, BW}

b. Each outcome is equally likely.

Result	Probability
0 B	4/9
1 B	4/9
2 B	1/9

69. There are $28 \times 4 = 112$ combinations.

71. There are $7 \times 9 \times 11 = 693$ systems.

73. a. There are 48 men at the convention, so the probability a random delegate is a man is $48/100 = 0.48$.

b. There are 41 Democrats, so the probability is $41/100 = 0.41$.

c. There are 10 Independents, and thus 90 non- Independents, which means the probability is $90/100 = 0.90$.

d. There are 28 male Republicans, so the probability is $28/100 = 0.28$.

e. There are 21 female Republicans, so the probability you meet one is $21/100 = 0.21$. Thus, the probability that you meet someone who is not a female Republicans, is $1 - 0.21 = 0.79$.

75. a. The probability, as shown in the table, is 51.2%, or 0.512. The table already *is* a probability distribution: it shows the probabilities (expressed as percents) that a randomly selected person will be in each of the categories.

b. $0.348 + 0.089 + 0.026 = 0.463$

c. No, unless you also know the total numbers of men and women in the population.

77. Answers will vary.

TECHNOLOGY EXERCISES

85. Answers will vary.

UNIT 7B: COMBINING PROBABILITIES

QUICK QUIZ

1. c. The probability of getting a double-1 on at least one of the rolls is $1 - (35/36)^2$.

2. c. The rule works only when A and B are independent events.

3. b. The probability of choosing a red M&M on the second pick is dependent upon what happens on the first pick (because there will be fewer M&Ms in the bag).

4. c. Since these are dependent events, the rule $P(A \text{ and } B \text{ and } C) = P(A) \times P(B \text{ given } A) \times P(C \text{ given } A \text{ and } B)$ must be used.

5. **c.** A person can be born on both a Monday and in June, so the events are overlapping.

6. **c.** $P(5 \text{ or } 6 \text{ or } 7 \text{ or } 8) = P(5)+P(6)+P(7)+P(8) = 4/36+5/36+6/36+5/36 = 20/36 > 0.5$

7. **a.** Since these events are independent, we have $P(\text{sum of 3 and sum of 4}) = P(\text{sum of 3}) \times P(\text{sum of 4})$ $= 2/36 \times 3/36$.

8. **b.** The probability of getting at least one success in ten trials of the experiment is $1-[P(\text{not two heads})]^{10}$, so you need first to compute the probability of not two heads.

9. **c.** This answer shows the correct application of the *at least once* rule.

10. **b.** This answer shows the correct application of the *at least once* rule.

REVIEW QUESTIONS

1. To understand his losses in his con games, the Chevalier asked mathematician Blaise Pascal, who, in turn, consulted with the mathematician Pierre de Fermat, to delve into the chances behind the outcomes of the games.

3. If you are drawing cards, you might be interested in drawing a red card, where the card is either a heart *or* a diamond. Two events are non-overlapping if they cannot occur together. Hearts and diamonds do not overlap, so $P(\text{red card}) = P(\text{heart}) + P(\text{diamond})$. Two events are overlapping if they can occur together. Red cards and Aces overlap, so $P(\text{red or Ace}) = P(\text{red}) + P(\text{Ace}) - P(\text{red Ace})$.

DOES IT MAKE SENSE?

5. Makes sense. You can't get both heads and tails on a single flip of a coin, but you must get one or the other.

7. Does not make sense. The probability of drawing an ace or a spade is $16/52$, but the probability of drawing the ace of spades is $1/52$.

9. Does not make sense. The chance of getting at least one 5 is $1-(5/6)^6 \neq 6/6 = 1$.

BASIC SKILLS AND CONCEPTS

11. Results will vary. You will not always get at least one head in your two rolls.

13. The events are independent.
$P(H \text{ and } H \text{ and } H \text{ and } H \text{ and } H) = P(H)\times P(H)\times P(H)\times P(H)\times P(H) = (1/2)^5 = 1/32$

15. The events are independent. $P(1 \text{ and } 2 \text{ and } 3) = P(1)\times P(2)\times P(3) = (1/6)^3 = 1/216$

17. The events are independent. $P(\text{1st and 2nd and 3rd}) = P(\text{1st})\times P(\text{2nd})\times P(\text{3rd}) = (1/7)^3 = 1/343$

19. The events are dependent, and the probability is $P(\text{1st red and 2nd red and 3rd red})$ $= (10/30)\times(9/29)\times(8/28) = 6/203$

21. The events are dependent, the probability is $P(\text{1st male and 2nd male and 3rd male and 4th male and 5th male})$ $= (10/20)\times(9/19)\times(8/18)\times(7/17)\times(6/16) = 21/1292$

23. The events are non-overlapping.
$P(2 \text{ or } 3 \text{ or } 4 \text{ or } 5) = P(2)+P(3)+P(4)+P(5) = 1/36+2/36+3/36+4/36 = 5/18$

25. The events are non-overlapping. $P(\text{black ace or red king}) = P(\text{black ace}) + P(\text{red king}) = 2/52+2/52 = 1/13$

27. The events overlap. $P(\text{black or small}) = P(\text{black}) + P(\text{small}) - P(\text{black and small}) = 1/2+1/2-1/4 = 3/4$

29. $P(\text{at least one head}) = 1-(1/2)^3 = 7/8 = 0.875$

31. $P(\text{rain at least once}) = 1-(0.7)^3 \approx 0.657$

33. $P(\text{at least one win}) = 1-(0.99)^{15} = 91/216 \approx 0.140$

FURTHER APPLICATIONS

35. The events are independent. The probability of choosing either die is $1/2$ and rolling an even result is $1/2$ for each die $P(\text{green even number}) = (1/2)\times(1/2) = 1/4$

37. The probability of not getting an even number on a single roll is 1/2, so $P(\text{at least one even}) = 1-(1/2)^5 = 31/32 = 0.96875$.

39. The events are dependent. $P(\text{four spades}) = (13/52)\times(12/51)\times(11/50)\times(10/49) \approx 0.00264$

41. The events are overlapping. $P(\text{face card or club}) = P(\text{face card}) + P(\text{club}) - P(\text{face card and club})$
$= 12/52 + 13/52 - 3/52 = 22/52 = 11/26 \approx 0.4231$

43. $P(\text{at least one green light}) = 1-(6/10)^5 \approx 0.922$

45. $P(\text{at least one ace}) = 1-(48/52)^6 \approx 0.381$

47. The events are independent. The probability of all five tickets being winners is $(1/8)^5 = 1/32{,}769 \approx 0.0000305$.

49. $P(\text{at least one left-handed person}) = 1-(0.89)^8 \approx 0.606$

51. The events are overlapping. $P(W \text{ or } R) = P(W) + P(R) - P(W \text{ and } R) = 110/160 + 80/160 - 60/160 = 130/160 = 13/16 = 0.8125$.

53. The events are non-overlapping. The easiest way to solve the problem is to look at the set of possible outcomes for a three child family, {GGG, GGB, GBG, BGG, BBG, BGB, GBB, BBB}, and count the number of outcomes with either all boys or all girls. The probability is $2/8 = 1/4$. One could also use the formulas for *or* probabilities.

55. $P(\text{at least one false test}) = 1-(0.98)^4 \approx 0.0776$

57. a. $P(G \text{ or } P) = P(G) + P(P) - P(G \text{ and } P) = 956/1028 + 450/1028 - 392/1028 \approx 0.986$

b. $P(NG \text{ or } NP) = P(NG) + P(NP) - P(NG \text{ and } NP) = 72/1028 + 578/1028 - 14/1028 \approx 0.619$

59. a. These events are dependent because there are fewer people to choose from on the second and subsequent calls, and this affects the probability that a Republican is chosen.

b. $P(R_1 \text{ and } R_2) = P(R_1) * P(R_2 \text{ given } R_1) = (25/45)\times(24/44) \approx 0.303$

c. If the events are treated as independent, then the pollster is selecting a person from random and not crossing that person off the list of candidates for the second call. The probability of selecting two Republicans is then $P(R_1 \text{ and } R_2) = P(R_1)\times P(R_2) = (25/45)\times(25/45) \approx 0.309$

d. The probabilities are nearly the same. This is due to the fact that the list of names is large, and removing one person from the list before making the second call doesn't appreciably change the probability that a Republican will be chosen.

61. The probability of winning would be $1-(35/36)^{25} \approx 0.506$. Since this is larger than 50%, he should come out ahead over time.

63. a. The probability is 1/20.

b. $P(\text{H}_1 \text{ and } \text{H}_2) = P(\text{H}_1)\times P(\text{H}_3) = (1/20)\times(1/20) = 1/400 = 0.0025$

c. $P(\text{at least one hurricane}) = 1-(19/20)^{10} \approx 0.401$

65. a. $P(\text{misses 1st free throw}) = 1 - 0.7 = 0.3$

  b. $P(\text{makes 1st and misses 2nd}) = (0.7)(0.3) = 0.21$

  c. $P(\text{makes 1st and makes 2nd}) = (0.7)(0.7) = 0.49$

  Making both free throws is most likely.

UNIT 7C: THE LAW OF LARGE NUMBERS

QUICK QUIZ

1. **b.** The law of large numbers says that the proportion of years should be close to the probability of a hurricane on a single year (and $100/1000 = 0.1$
2. **c.** The expected value is $(-\$1) \times 1 + \$75 \text{ million} \times 1/100{,}000{,}000 = -\$0.25$.
3. **c.** If you purchase 1/10 of the available number of lottery tickets, there's a 1 in 10 chance that you have the winning ticket. But this means that 90% of the time you have purchased 1 million losing tickets.
4. **c.** The expected value can be understood as the average value of the game over many trials.
5. **b.** The expected value of a \$2000 fire insurance policy from the point of view of the insurance company is $\$2000 \times 1 + (-\$100{,}000) \times 1/50 = 0$, so they can expect to lose money if they charge less than \$2000 per policy.
6. **a.** The shortcut saves 5 minutes 90% of the time, but it loses 20 minutes 10% of the time.
7. **c.** The results of the previous ten games have no bearing on the outcome of the next game.
8. **c.** The results of the previous ten games have no bearing on the next game.
9. **c.** In order to compute the probability of winning on any single trial, you would need to know the probability of winning on each of the various payouts.
10. **a.** 97% of \$10 million is \$9.7 million, which leaves \$300,000 in profit for the casino (over the long haul).

REVIEW QUESTIONS

1. Consider an event A with probability $P(A)$ in a single trial. The law of large numbers holds that: For a large number of trials, the proportion in which event A occurs will be close to the probability $P(A)$. The larger the number of trials, the closer the proportion should be to $P(A)$. This law holds as long as each trial is independent of prior trials, so that an individual trial always has the same probability, $P(A)$. The law of large numbers does not apply to single experiments as it only apply to long term average probabilities.

3. Expected value is the average result for a long series of events. It is calculated by finding the sum of each outcome times its probability of occurring. Since it is an average, the expected value may not represent a possible outcome. For example, the expected value of many rolls of a die is 3.5, which is not a possible outcome of a die roll.

5. Coin tosses can be assumed to be independent, so the probability of tossing a head or tail is the same for each toss. The probability of any set of outcomes is $(\frac{1}{2})\times(\frac{1}{2})\times(\frac{1}{2})\times(\frac{1}{2})\times(\frac{1}{2})\times(\frac{1}{2})\times(\frac{1}{2})\times(\frac{1}{2})\times(\frac{1}{2})\times(\frac{1}{2}) = (1/2)^{10}$. The probability of tossing a head or tail on the next toss is still $\frac{1}{2}$.

DOES IT MAKE SENSE?

7. Makes sense. An organization holding a raffle usually wants to raise money, which means the expected value for purchasers of raffle tickets will be negative.

9. Does not make sense. Each of the 16 possible outcomes when four coins are tossed is equally likely.

11. Does not make sense. The slot machine doesn't remember what happened on the previous 25 pulls, and your probability of winning remains the same for each pull.

BASIC SKILLS AND CONCEPTS

13. You should not expect to get exactly 5000 heads (the probability of that happening is quite small), but you should expect to get a result near to 5000 heads. The law of large numbers says that the proportion of a large number of coin tosses that result in heads is near to the probability of getting heads on a single toss. Since the probability of getting heads is $1/2$, you can expect that around one-half of the 10,000 tosses will result in heads.

15. The probability of tossing three heads is 1/8, and that of not tossing three heads is 7/8. Thus the expected value of the game is $\$5\times1/8+(-\$1)\times7/8=-\$0.25$. Though one can assign a probability to the outcome of a single game, the result cannot be predicted, so you cannot be sure whether you will win or lose money in one game. However, over the course of 100 games, the law of averages comes into play, and you can expect to lose about $25.

17. The probability of rolling two even numbers is $9/36=1/4$ (of the 36 possible outcomes, 9 are "two even numbers"), and the probability of not rolling two even numbers is $27/36=3/4$. Thus the expected value is $\$3\times1/4+(-\$1)\times3/4=\$0$. The outcome of one game cannot be predicted, though over 100 games, you should expect to break even.

19. The expected value for a single policy is $\$300+(-\$10{,}000)\times1/100+(-\$25{,}000)\times1/250$ $+(-\$50{,}000)\times1/500=\$0$. The company can expect a profit of $\$0\times10{,}000=\$0$ if they sell 10,000 policies.

21. For the sake of simplicity, assume that you arrive on the half-minute for each minute of the hour (that is, you arrive 30 seconds into the first minute of the hour, 30 seconds into the second minute of the hour, etc.). The probability that you arrive in any given minute in the first 20 minutes is 1/20, and your wait times (in minutes) for the first 20 minutes are 19.5, 18.5, 18.5, … , 0.5. Thus your expected wait time is $19.5\times1/20+18.5\times1/20$ $+17.5\times1/20+\ldots+1.5\times1/20+0.5\times1/20\approx10$ minutes. Of course the same is true for the second and third 20 minute intervals. (Note for instructors: if you wanted to allow for arrival at *any* time during a given minute, you would need to use calculus in order to solve the problem with any level of rigor, because the wait time function is continuous.)

23. a. Heads came up 46% of the time, and you've lost $8.

    b. Yes, the increase in heads is consistent with the law of large numbers: the more often the coin is tossed, the more likely the percentage of heads is closer to 50%. Since 47% of 300 is 141, there have been 141 heads and 159 tails, and you are now behind $18.

    c. You would need 59 heads in the next 100 tosses (so that the total number of heads would be 200) in order to break even. While it's possible to get 59 heads in 100 tosses, a number nearer to 50 is more likely.

    d. If you decide to keep playing, the most likely outcome is that half of the time you'll see heads, and half the time you'll see tails, which means you'll still be behind, overall.

25. a. If you toss a head, the difference becomes 15; if you toss a tail, the difference becomes 17.

    b. The most likely outcome after 1000 more flips is 500 heads and 500 tails (in which case the difference would remain at 16). The distribution of the number of heads is symmetric so that any deviation from "500 heads/500 tails" is as likely to be in the upper tail as the lower tail.

    c. Once you have fewer heads than tails, the most likely occurrence after additional coin tosses is that the deficit of heads remains.

27. Both sets are equally likely as each is one of many possible outcomes when six numbers are drawn.

29. The chance of hitting five free throws in a row is about $(7/10)^5=0.17$, so by chance Maria could have such a game in a 25-game season.

31. a. Suppose you bet \$1. Then your expected value is $\$1\times 0.493+(-\$1)\times 0.507=-\$0.014$. In general, the expected value is –1.4% of your bet, which means the house edge is 1.4% (or 1.4¢ per dollar bet).
    b. You should expect to lose $100\times \$0.014=\$1.40$.
    c. You should expect to lose $\$500\times 1.4\%=\$7$.
    d. The casino should expect to earn 1.4% of \$1 million, which is \$14,000.

FURTHER APPLICATIONS

33. The expected value is $(-\$1)\times 1+\$30{,}000{,}000\times 1/292{,}201{,}338+\$1{,}000{,}000\times 1/11{,}688{,}053$ $+\$50000\times 1/913{,}129+\$100\times 1/36{,}525+\$100\times 1/14{,}494+\$7\times 1/580+\$7\times 1/701+\$4\times 1/92+\$4\times 1/38$ $=-\$0.58$. You can expect to lose $52\times 10\times \$0.58=\$300$ over the year.

35. The expected value is $(-\$1)\times 1+\$15{,}000{,}000\times 1/258{,}890{,}850+\$1{,}000{,}000\times 1/18{,}492{,}203$ $+\$5000\times 1/739{,}688+\$500\times 1/52{,}835+\$5\times 1/766+\$5\times 1/473+\$2\times 1/56+\$1\times 1/21=-\$0.77$. You can expect to lose $100\times \$0.77=\$77$ if you bought 100 lottery tickets.

37. a. $23{,}325/23{,}684\approx 0.9848$; $1\times 0.9848=0.9848$
    b. $2265/2412\approx 0.9391$; $1\times 0.9391=0.9391$
    c. $344/718\approx 0.4791$; $2\times 0.4791=0.9582$
    d. Before the kicking distance was increased, one-point conversions were best, with an expected value of 0.9848. After the change, two-point conversions are best, with an expected value of 0.9582.

39. The expected value for the number of people in a household is $1.5\times 0.57+3.5\times 0.32+6\times 0.11=2.6$ people. The expected value is a weighted mean, and it is a good approximation to the actual mean as long as the midpoints are representative of the categories.

UNIT 7D: ASSESSING RISK

QUICK QUIZ

1. b. $40{,}000/365\approx 109$
2. c. The blue curve in the graph is rising as time progresses.
3. b. The death rate per vehicle-mile has generally decreased with time.
4. c. The death rate per person is the number of deaths due to AIDS divided by the population.
5. c. The death rate per person is $\frac{76{,}500}{325 \text{ million}}$, and this needs to be multiplied by 100,000 to get the death rate per 100,000 people.
6. a. The chart shows that children in their younger years have a significant risk of death.
7. c. Your life expectancy decreases as you age because the number of years before you die is decreasing.
8. c. Current life expectancies are calculated for current conditions – if these change, life expectancies can also change.
9. b. The life expectancies of men are consistently below those of women (and from the graph, the gap has widened a bit in the last several decades).
10. a. If life expectancy goes up, people will live longer, and will draw Social Security benefits for a longer period of time.

REVIEW QUESTIONS

1. Quantifying risk allows us to make informed decisions about the benefits and dangers of a decision.
3. Vital statistics consist of data concerning the births and deaths of people. They include birth and death rates and causes of death.

DOES IT MAKE SENSE?

5. Does not make sense. Automobiles are successfully sold in huge quantities despite the risks involved.

7. Does not make sense. Your life expectancy is an average life expectancy for people born when you were. Much more important is your genetic makeup, and the choices you make throughout life (your diet, your line of work, the activities you choose to participate in, etc.).

BASIC SKILLS AND CONCEPTS

9. Killed: $\frac{1\text{ year}}{40{,}200\text{ fatalities}}\times\frac{365\text{ days}}{1\text{ year}}\times\frac{24\text{ hours}}{1\text{ day}}\times\frac{60\text{ min}}{1\text{ hour}}\approx 13\text{ minutes/fatality}$

Injured: $\frac{1\text{ year}}{4{,}600{,}000\text{ injuries}}\times\frac{365\text{ days}}{1\text{ year}}\times\frac{24\text{ hours}}{1\text{ day}}\times\frac{60\text{ min}}{1\text{ hour}}\times\frac{60\text{ sec}}{1\text{ min}}\approx 7\text{ seconds/injury}$

11. The per-person cost is $\frac{\$430\times 10^9}{325\times 10^6}\approx \$1320/\text{person}$.

13. a. In 1995, the per-person fatality rate was $\frac{41{,}817}{2.63\times 10^8}=1.59\times 10^{-4}$. Multiply this by 100,000 to get 15.9 fatalities per 100,000 people. There were 23.6 fatalities per 100,000 drivers, and 1.7 fatalities per 100 million vehicle miles (the calculations are similar). In 2015, these figures were 10.6, 15.7, and 1.1, respectively. In all three measures, the death rates declined from 1995 to 2010.

a. 1995: $\frac{41{,}817}{2.4\times 10^{12}}\times 100\times 10^6=1.7$ fatalities per 100 million miles

2015: $\frac{35{,}092}{3.1\times 10^{12}}\times 100\times 10^6=1.1$ fatalities per 100 million miles

b. 1995: $\frac{41{,}817}{263\times 10^6}\times 10^5=15.9$ fatalities per 100,000 people

2015: $\frac{35{,}092}{321\times 10^6}\times 10^5=10.9$ fatalities per 100,000 people

c. 1995: $\frac{41{,}817}{177\times 10^6}\times 10^5=23.6$ fatalities per 100,000 licensed drivers

2015: $\frac{35{,}092}{218\times 10^6}\times 10^5=16.1$ fatalities per 100,000 licensed drivers

d. They are generally consistent, as all three measures declined by similar amounts from 1995 to 2015. (The decline per vehicle-mile was slightly greater than the declines per person or per driver.)

15. a. In 2000 there were $\frac{49}{16.7\times 10^6}\times 10^6=2.9$ accidents per 1 million hours flown. In 2015 there were $\frac{27}{17.4\times 10^6}\times 10^6=1.6$ accidents per 1 million hours flown. The rates are close, but it seems flying was a little less safe in 2000.

b. In 2000 there were $\frac{49}{7.1\times 10^9}\times 10^9=6.9$ accidents per 1 billion miles flown. In 2015 there were $\frac{27}{7.6\times 10^9}\times 10^9=3.6$ accidents per 1 billion miles flown. It seems flying was less safe in 2000.

17. The probability of death by diabetes is $\frac{76{,}500}{3.25\times10^8} = 0.00024$. The probability of death by kidney disease is $\frac{48{,}100}{3.25\times10^8} = 0.00015$. Death by diabetes is about $\frac{0.0002}{0.00015} = 1.6$ times more likely than death by kidney disease.

19. The death rate due to stroke is about $\frac{133{,}100}{3.25\times10^8}\times10^5 = 40.95$ deaths per 100,000 people, so in a city of 500,000, you would expect $40.95\times5 \approx 205$ people to die from a stroke each year.

21. The death rate for the category of 60- to 65-year-olds is about 11 deaths per 1000 people. Multiply 11/1000 by 14.2 million to find the number of people in this category expected to die in a year: $\frac{11}{1000}\times14.2\times10^6 \approx 156{,}200$ people. (Note: Answers will vary depending on values read from the graphs)

23. The death rate for a 50-year-old is about 5 per 1000, which means the probability that a randomly selected 50-year-old will die in the next year is 5/1000. Those that do not die will file no claim; the beneficiaries of those that do die will file a $50,000 claim. Thus the expected value of a single policy (from the point of view of the insurance company) is $\$200\times1+(-\$50{,}000)\times(5/1000)+(\$0)\times(995/1000) = -\$50$. The company insures 1 million 50-year-olds, and expects to lose $50 per policy, so it can expect a loss of $50 million.

25. Life expectancy for women increased by $\frac{80-48}{48} = 0.6667 = 66.67\%$, so the life expectancy for women in 2100 would be $80\times1.6667 = 133$ years. Answers will vary, but it appears the prediction in Example 5 seems more reasonable.

FURTHER APPLICATIONS

27. a. In Utah, about $\frac{51{,}154}{365} = 140$ people were born per day.

b. In New Hampshire, about $\frac{12{,}302}{365} = 34$ people were born per day.

c. The birth rate in Utah was $\frac{51{,}154}{3.0\times10^6}\times10^5 = 1705$ births per 100,000 people.

d. The birth rate in New Hampshire was $\frac{12{,}302}{1.3\times10^6}\times10^5 = 946$ births per 100,000 people.

29. a. There were about $\frac{12.5}{1000}\times321$ million $= 4.01$ million births in the U.S.

b. There were about $\frac{8.4}{1000}\times321$ million $= 2.70$ million deaths in the U.S.

c. The population increased by 4.01 million $- 2.70$ million $= 1.31$ million people.

d. About 3.0 million $- 1.31$ million $= 1.7$ million people immigrated to the U.S. (This estimate neglects to account for emigration, though the number of people who emigrate from the U.S. is relatively low.) The proportion of the overall population growth due to immigration was $1.7/3.0 = 57\%$.

31. a.

Year	Percentage of Americans over age 65
1950	$\frac{12.7}{151}\times 100 = 8.4\%$
1960	$\frac{17.2}{179}\times 100 = 9.6\%$
1970	$\frac{20.9}{203}\times 100 = 10.3\%$
1980	$\frac{26.1}{227}\times 100 = 11.5\%$
1990	$\frac{31.9}{249}\times 100 = 12.8\%$
2000	$\frac{34.9}{281}\times 100 = 12.4\%$
2010	$\frac{40.4}{309}\times 100 = 13.1\%$
2015	$\frac{47.7}{321}\times 100 = 14.9\%$

b. $\frac{321-151}{151}\times 100 = 113\%$

c. $\frac{47.7-12.7}{12.7}\times 100 = 276\%$

d. Assuming half of 25–65 aged Americans work, 84.5 million contribute, which is roughly twice the over 65 population.

e. Answers will vary.

UNIT 7E: COUNTING AND PROBABILITY

QUICK QUIZ

1. **c. This is an arrangement with repetition, with 36 choices at each selection.**
2. **c. Assuming each person gets one entrée, this is a permutation of 4 items taken 4 at a time, and $_4P_4 = 4!/0! = 24$.**
3. **c. The order of selection is important when counting the number of ways to arrange the roles.**
4. **b. The permutation of 12 items taken 5 at a time is $_{12}P_5$.**
5. **b. The number of permutation of n items taken r at a time is always at least as large as the number of combinations of n items taken r at a time (by a factor of $r!$). and $_{15}P_7 = 32{,}432{,}400$ is much larger than $_7P_7 = 5040$.**
6. **c. The variable n stands for the number of items from which you are selecting, and the variable r stands for the number of items you select. Here, $n = 9$, and $r = 4$, and thus $(n-r)! = 5!$.**
7. **c. The number of combinations of 9 items taken 4 at a time is $_9C_4 = \frac{9!}{(9-4)!4!} = \frac{9\times 8\times 7\times 6}{4\times 3\times 2\times 1}$.**
8. **b. If there is a drawing, the probability that one person is selected is 100%.**

9. **c.** The probability that it will be you is $1/100{,}000 = 0.00001$, so the probability it will not be you is $1 - 0.00001 = 0.99999$.

10. **c.** As shown in Example 8b, the probability is about 57%.

REVIEW QUESTIONS

1. Arrangements with repetition are counting problems in which we repeatedly select from the same group of choices. Selecting a 6 digit ID number where numbers can be repeated would have 10^6 possible arrangements with repetition.

3. See the Information Box on page 479.

DOES IT MAKE SENSE?

5. Makes sense. The permutations formula is used when order of selection is important.

7. Makes sense. The number of such batting orders is ${}_{25}P_9 = 7.4 \times 10^{11}$, which is nearly 1 trillion ways. Clearly there is no hope of trying all of them.

9. Makes sense. This illustrates the general principle that *some* coincidence is far more likely that a *particular* coincidence.

BASIC SKILLS AND CONCEPTS

11. $6! = 720$

13. $\frac{5!}{3!} = 20$

15. $\frac{12!}{4!3!} = 3{,}326{,}400$

17. $\frac{11!}{3!(11-3)!} = 165$

19. $\frac{8!}{3!(8-3)!} = 56$

21. $\frac{6!8!}{4!5!} = 10{,}800$

23. This is an arrangement with repetition, where we have 10 choices (the digits 0–9) for each of 7 selections, and thus there are $10^7 = 10{,}000{,}000$ possible phone numbers.

25. This is a permutation of 26 items taken 5 at a time because the order in which the characters are arranged is important (a different order results in a different password). Thus there are ${}_{26}P_5 = 7{,}893{,}600$ passwords.

27. This is a permutation (of 8 items taken 4 at a time) because once the four committee members are chosen, the offices which they are to hold must be determined (and different orders produce different people in the various offices). The number of such executive committees is ${}_8P_4 = 1680$.

29. Since every candidate can be assigned to any position, the order of selection matters (it determines the assigned country, for example), and this is a permutation of 10 items taken 6 at a time. The number of such permutations is ${}_{10}P_6 = 151{,}200$.

31. Because order of selection does not matter, this is a combination of 12 items taken 6 at a time, or ${}_{12}C_6 = 924$.

33. Since every player can be assigned to any position, the order of selection matters (perhaps it determines the position played, for example), and this is a permutation of 10 items taken 6 at a time. The number of such permutations is ${}_{10}P_6 = 151{,}200$.

35. This is an arrangement with repetition, so use the multiplication principle to find that there are $26^3 \times 10^4 = 175{,}760{,}000$ such license plates.

37. This is an arrangement with repetition, where there are 2 choices for each of 6 selections, and thus there are $2^6 = 64$ birth orders.

39. If we assume that the letters can be repeated, this is an arrangement with repetition, and there are 4 choices for each of the 3 selections, which implies there are $4^3 = 64$ different "words."

41. Following Example 8 in the text, the probability that no one has your birthday is $1-\left(\frac{364}{365}\right)^{11} \approx 0.0297$. The probability that at least one pair shares a birthday is given by $1-\frac{364\times 363\times \ldots \times 354}{365^{11}} = 1-\frac{1.276\times 10^{28}}{1.532\times 10^{28}}$ ≈ 0.167.

FURTHER APPLICATIONS

43. a. Use the multiplication principle to get $20\times 8 = 160$ different sundaes.

b. This is an arrangement with repetition, so there are $20^3 = 8000$ possible triple cones.

c. Since you specify the locations of the flavors, the order matters, and you should use permutations. There are ${}_{20}P_3 = 6840$ possibilities.

d. The order of the flavors doesn't matter, so use combinations. There are ${}_{20}C_3 = 1140$ possibilities.

45. This problem is best solved by trial and error. For Luigi, we know that 56 is the number of combinations of *n* items taken 3 at a time, and thus ${}_nC_3 = 56$. After a little experimentation, you will discover that $n = 8$; that is, Luigi uses 8 different toppings to create 56 different 3-topping pizzas. In a similar fashion, you can discover that Ramona uses 9 different toppings (solve ${}_nC_2 = 36$ for *n*).

47. There are ${}_{32}C_6 = 906{,}192$ ways to choose the balls, and only one way to match all six numbers, so the probability is $1/906{,}192$.

49. There are ${}_{52}C_5 = 2{,}598{,}960$ five-card hands. Four of those consist of the 10, J, Q, K, and A of the same suit, so the probability $4/2{,}598{,}960 = 1/649{,}740$.

51. There are ${}_{12}C_4 = 495$ ways to choose the top four spellers, and only one way to guess the correct four, so the probability is $1/495$.

53. There are ${}_{52}C_5 = 2{,}598{,}960$ five-card hands. In order to get four-of-a-kind in, say, aces, you need to select all four aces, and then any of the other remaining 48 cards. Thus there are 48 ways to get a four-of-a-kind in aces. This is true for the other 12 card values as well, so there are $13\times 48 = 624$ different four-of-a-kind hands. The desired probability is then $624/2{,}598{,}960 \approx 0.00024$.

55. a. The probability of any one individual winning five games in a row is $0.48^5 = 0.025$, assuming you play only five games. It is considerably higher if you play numerous games. If 2000 people each play 5 games in a row, we would expect about $2000\times 0.025 = 50$ people to have such a "hot streak."

b. The probability of any one individual winning ten games in a row is $0.48^{10} = 0.00065$, assuming you play only ten games. If 2000 people each play 10 games in a row, we would expect about $2000\times 0.00065 = 1.3$ people to have such a "hot streak."

57. There are ${}_{75}C_5 = 17{,}259{,}390$ ways to choose the first five numbers and 15 ways to choose Megaball number, so there are $17{,}259{,}390\times 15 = 258{,}890{,}850$ ways to play. There is only one way to win the jackpot, so the probability is $1/258{,}890{,}850$ or about 0.000000004.

TECHNOLOGY EXERCISES

63. a. There are ${}_{44}C_5 = 1{,}086{,}008$ ways to choose the balls, and only one way to match all five numbers, so the probability is $1/1{,}086{,}008$.

    b. There are ${}_{40}C_6 = 3{,}838{,}380$ ways to choose the balls, and only one way to match all six numbers, so the probability is $1/3{,}838{,}380$.

    c. The first lottery offers a better chance of winning.

65. Results will vary.

UNIT 8A: GROWTH: LINEAR VS. EXPONENTIAL

QUICK QUIZ

1. **b.** The absolute change in the population was 20,000 people. If population grows linearly, the absolute change remains constant, thus the population at the end of the second year will be $120,000 + 20,000 = 140,000$.

2. **c.** Because the population increases exponentially, it undergoes constant percent change. Its population increases by 20% in the first year, and thus at the end of the second year, the population will be $120,000 \times 20\% + 120,000 = 144,000$.

3. **b.** Because your money is growing exponentially, if it doubles in the first 6 months, it will double every 6 months.

4. **b.** The absolute change in the number of songs is 200, and this occurred over the span of 3 months. Because the number of songs is increasing linearly, it will grow by 200 every 3 months, and it will take 6 months to grow to 800 songs.

5. **c.** Exponential decay occurs whenever a quantity decreases by a constant percentage (over the same time interval).

6. **c.** Based on the results of Parable 1 in the text, the total number of pennies needed is $2^{64} - 1 = 1.8 \times 10^{19}$, which is equivalent to $\$1.8 \times 10^{17}$. The federal debt is about \$20 trillion $\left(\$20 \times 10^{12}\right)$, which differs from the amount of money necessary by about a factor of 10,000.

7. **b.** Note that at 11:01, there are $2 = 2^1$ bacteria; at 11:02 there are $4 = 2^2$ bacteria; at 11:03 there are $8 = 2^3$, and so on. The exponent on the base of 2 matches the minute, and this pattern continues through 11:30, which means there are 2^{30} bacteria at that time.

8. **b.** At 11:31, there are 2^{31} bacteria, and as shown in Exercise 7, there were 2^{30} bacteria at 11:30. The difference, $2^{31} - 2^{30} = 2^{30}$, which is the number of bacteria added over that minute.

9. **a.** As with the population, the volume of the colony doubles every minute, so it only takes one minute for the volume to double from 1 to 2 m^3.

10. **b.** The growth of a population undergoing exponential change is comparatively slow in the initial stages, and then outrageously fast in the latter stages. This makes it difficult for members of the population to see that the growth they are currently experiencing is about to explode in short order.

REVIEW QUESTIONS

1. In linear growth, the given quantity grows by the same *absolute* amount each fixed time period. In exponential growth, the given quantity grows by the same *percentage* every fixed time period

3. The bacteria in a bottle double their population every minute, leading to disaster as the bottle fills to capacity.
Question 1: The bottle was half-full at 11:59, just 1 minute before the disaster.
Question 2: You would not have enough time, that the bottle remains nearly empty for most of the 60 minutes, but the continued doublings fill it rapidly in the final 4 minutes.
Question 3: The discovery of three new bottles gives the bacteria only 2 additional minutes.
Question 4: If the doublings continued for just $5\frac{1}{2}$ hours, the volume of bacteria would exceed the volume of the entire universe

DOES IT MAKE SENSE?

5. Makes sense. Any quantity that undergoes constant, positive percent change grows exponentially.

7. Makes sense. Any quantity that grows exponentially will eventually become very large, so as long as the small town's rate of growth is large enough, it can become a large city within a few decades.

BASIC SKILLS AND CONCEPTS

9. The population is growing linearly because its absolute change remains constant at 300 people per year. The population in four years will be $2400 + 4 \times 300 = 3600$.

11. This is exponential growth because the cost of food is increasing at a constant rate of 40% per month. Each month, the cost increases by a factor of 1.40, so your food bill would be 1000 bolivars $\times 1.40^4 = 3841.60$ bolivars.

13. The price is decreasing exponentially because the percent change remains at a constant -15% per year. Each year, the price decreases by a factor of 0.85, so its price in three years will be $\$50(0.85)^3 = \$31.71$.

15. The house's value is increasing linearly because the price increases by the same amount each year. In five years, it will be worth $\$100{,}000 + 5 \times \$2{,}000 = \$110{,}000$.

17. There should be $2^{15} = 32{,}768$ grains placed on the 16th square. The total number of grains would be $2^{16} - 1 = 65{,}535$, and it would weigh $65{,}535 \times (1/7000) = 9.36$ pounds.

19. When the chessboard is full, there are $2^{64} - 1$ grains, with total weight of $\left(2^{64} - 1\right) \times (1/7000) = 2.64 \times 10^{15}$ pounds, or about 1.3×10^{12} tons.

21. After 21 days, you would have $2^{21} = 2{,}097{,}152$ pennies, or \$20,971.52.

23. After experimenting with various values of n in 2^n, you'll find that the balance has grown to about \$1.4 billion after 37 days.

25. At 11:50, there are $2^{50} = 1.1 \times 10^{15}$ bacteria in the bottle. The bottle is full at 12:00 noon, so it was half-full at 11:59, 1/4-full at 11:58, 1/8-full at 11:57, and so on. Continuing in this manner, we find the bottle was 1/1024-full at 11:50. This could also be expressed as $1/2^{10}$-full at 11:50.

27. As discussed in the text, the volume of the colony of 2^{120} bacteria would occupy a volume of $1.3 \times 10^{15}\ \text{m}^3$. Divide this by the surface area of the earth to get an approximate depth: $1.3 \times 10^{15}\ \text{m}^3 \div 5.1 \times 10^{14}\ \text{m}^2 = 2.5$ m. This is quite a bit more than knee-deep (it's more than 8 feet).

FURTHER APPLICATIONS

29. a. (Only 100-year intervals are shown here).

Year	Population
2000	6.0×10^9
2100	2.4×10^{10}
2200	9.6×10^{10}
2300	3.8×10^{11}
2400	1.5×10^{12}
2500	6.1×10^{12}

Year	Population
2600	2.5×10^{13}
2700	9.8×10^{13}
2800	3.9×10^{14}
2900	1.6×10^{15}
3000	6.3×10^{15}

b. In the year 2800, there would be 3.9×10^{14} people, each occupying slightly more than one square meter, and in 2850, there would be 7.9×10^{14} people (each occupying slightly less than one square meter). Since $5.1 \times 10^{14}\ \text{m}^2$ is in between these two values, it would happen between 2800 and 2850.

29. (continued)

c. We would reach the limit when the earth's population reached $5.1\times10^{14}\div10^{4}=5.1\times10^{10}$ people. This would happen shortly after the year 2150.

d. If by 2150 we've already reached the limit of $10^4\ \text{m}^2$ per person on earth, it will take only three more doubling periods to increase our need for space eight-fold. Since the doubling period is 50 years, this would take only 150 years, at which point we would have already exceeded the additional space available in the solar system (there is only five times more area out there).

31. a. and b.

Month	December 2013	December 2014	December 2015	December 2016
Monthly Active users (millions)	1228	1393	1591	1860
Absolute change	–	$1393-1228=165$	$1591-1393=198$	$1860-1591=269$
Percent change	–	$\frac{1393-1228}{1228}=13.4\%$	$\frac{1591-1393}{1393}=14.2\%$	$\frac{1860-1591}{1591}=16.9$

c. The absolute change is roughly constant and the percent change is decreasing, so the growth is closer to linear.

UNIT 8B: DOUBLING TIME AND HALF-LIFE

QUICK QUIZ

1. **a.** The factor by which an exponentially growing quantity grows is $2^{t/T_{\text{double}}}$.
2. **b.** The approximate doubling time formula says that a quantity's doubling time is about $70/r$, where r is the growth rate.
3. **a.** The approximate doubling time formula does not do a good job when the growth rate r goes higher than 15%.
4. **b.** The approximate doubling time formula, $T_{\text{double}}=70/r$, can be rearranged to say $r=70/T_{\text{double}}$.
5. **c.** During the first 12 years, half of the tritium decays, and during the second 12 years, half of that, or 1/4 decays.
6. **b.** Note that 2.8 billion years/4 = 700 million years. This implies the uranium-235 decayed by a factor of 2 four times, which means $1/2^4=1/16$ of the uranium-235 remains.
7. **a.** The half-life can be approximated by $T_{\text{half}}=70/r$.
8. **c.** A property of logarithms states $\log_{10}(10^x)=x$.
9. **c.** The decimal equivalent of 20% is 0.2, though you should set $r=-0.2$ because the population is decreasing.
10. **a.** The doubling time formula states $T_{\text{double}}=\dfrac{\log_{10}2}{\log_{10}(1+r)}$.

REVIEW QUESTIONS

1. Doubling time is the time required for each doubling in exponential growth to occur. In 25 years, the population would grow by a factor of $2^{25/25}=2^1=2$; in 50 years, by a factor of $2^{50/25}=2^2=4$; in 100 years, by a factor of $2^{100/25}=2^4=16$.

3. For a quantity growing exponentially at a rate of $P\%$ per time period, the doubling time is approximately $\frac{70}{P}$. This approximation works best for small growth rates and breaks down for growth rates over about 15%.

5. Multiply the initial population by $\left(\frac{1}{2}\right)^{t/T_{\text{half}}}$.

7. $T_{\text{double}} = \frac{\log_{10} 2}{\log_{10}(1+r)}$; where r is the fractional growth rate and $\log_{10}$ is the base 10 logarithm of the given input. The units of time for T_{double} and r must be the same.

   $T_{\text{half}} = -\frac{\log_{10} 2}{\log_{10}(1+r)}$; where r is the fractional growth rate and $\log_{10}$ is the base 10 logarithm of the given input. The units of time for T_{half} and r must be the same.

DOES IT MAKE SENSE?

9. Does not make sense. In 50 years, the population will increase by a factor of 4.

11. Does not make sense. While it's true that half will be gone in ten years, over the next ten years, only half of what remains will decay, which means there will still be 1/4 of the original amount.

BASIC SKILLS AND CONCEPTS

13. True. $10^0 = 1$, and $10^1 = 10$, so $10^{0.928}$ should be between 1 and 10 (because 0.928 is between 0 and 1).

15. False. $10^{-5.2} = \frac{1}{10^{5.2}}$, which is positive.

17. False. While it is true that π is between 3 and 4, $\log_{10} \pi$ is not (it's between 0 and 1).

19. False. $\log_{10}\left(10^6\right) = 6$, and $\log_{10}\left(10^7\right) = 7$. Because 1,600,000 is between 10^6 and 10^7, $\log_{10} 1{,}600{,}000$ is between 6 and 7.

21. True. Since $\log_{10}\left(\frac{1}{4}\right) = \log_{10}\left(4^{-1}\right) = -\log_{10} 4$, we know $\log_{10}\left(\frac{1}{4}\right)$ is between -1 and 0 because $\log_{10} 4$ is between 0 and 1.

23. a. $\log_{10} 16 = \log_{10}\left(2^4\right) = 4 \times \log_{10} 2 = 4 \times 0.301 = 1.204$

    b. $\log_{10} 20{,}000 = \log_{10} 2 + \log_{10} 10^4 = 0.301 + 4 = 4.301$

    c. $\log_{10} 0.05 = \log_{10}\left(\frac{1}{2} \times \frac{1}{10}\right) = \log_{10} 2^{-1} + \log_{10} 10^{-1} = -1 \times \log_{10} 2 + (-1) = -1 \times 0.301 - 1 = -1.301$

    d. $\log_{10} 128 = \log_{10} 2^7 = 7 \times \log_{10} 2 = 7 \times 0.301 = 2.107$

    e. $\log_{10} 0.02 = \log_{10}\left(\frac{2}{100}\right) = \log_{10} 2 - \log_{10} 100 = 0.301 - 2 = -1.699$

    f. $\log_{10}\left(\frac{1}{32}\right) = \log_{10} 2^{-5} = -5 \log_{10} 2 = -5 \times 0.301 = -1.505$

25. In 36 hours, the fly population will increase by a factor of $2^{36/12} = 2^3 = 8$. Since one week is 168 hours, the population will increase by a factor of $2^{168/12} = 2^{14} = 16{,}384$ in one week.

27. The population will increase by a factor of four (quadruple) in 40 years because $2^{40/20} = 2^2 = 4$.

29. In 12 years, the population will be $15{,}000\times2^{12/10}\approx34{,}000$. In 24 years, it will be $15{,}000\times2^{24/10}\approx79{,}000$.

31. In 20 months, the number of cells will be $1\times2^{20/1.5}=10{,}321$. In 3 years (36 months), the number will be $1\times2^{36/1.5}=16{,}777{,}216$.

33. The year 2027 is 10 years later than 2017, so the population will be $7.5\times2^{10/40}=8.9$ billion. In 2067 (50 years later), the population will be $7.5\times2^{50/40}=17.8$ billion. In 2117 (100 years later), it will be $7.5\times2^{100/40}=42.4$ billion.

35.

Month	Population
0	100
1	$100\times(1.07)^1=107$
2	$100\times(1.07)^2=114$
3	$100\times(1.07)^3=123$
4	$100\times(1.07)^4=131$
5	$100\times(1.07)^5=140$
6	$100\times(1.07)^6=150$
7	$100\times(1.07)^7=161$
8	$100\times(1.07)^8=172$
9	$100\times(1.07)^9=184$
10	$100\times(1.07)^{10}=197$
11	$100\times(1.07)^{11}=210$
12	$100\times(1.07)^{12}=225$
13	$100\times(1.07)^{13}=241$
14	$100\times(1.07)^{14}=258$
15	$100\times(1.07)^{15}=276$

Because the population after ten months is 197, which is almost twice the initial population of 100, the doubling time is just over 10 months. The approximate doubling time formula claims that $T_{\text{double}}\approx70/7=10$ months, which is pretty close to what we can discern from the table. Neither answer is exact, though.

37. $T_{\text{double}}\approx70/3.2=21.9$ years; Prices will increase by a factor of $(1.032)^3=1.10$ in three years. Using the doubling time formula, prices would increase by a factor of $2^{3/21.9}=1.10$, so the approximate doubling time formula does a reasonable job.

39. $T_{\text{double}}\approx70/1.2=58$ months; Prices will increase by a factor of $(1.012)^{12}=1.15$ in one year, and by a factor of $(1.012)^{96}=3.14$ in eight years. Using the doubling time formula, prices would increase by a factor of $2^{12/58}=1.15$ in one year, and by a factor of $2^{96/58}=3.15$ in eight years, so the approximate doubling time formula does a reasonable job.

41. After 80 years, the fraction of radioactive substance that remains is $\left(\frac{1}{2}\right)^{80/40}=\frac{1}{4}=0.25$. After 120 years, the fraction remaining is $\left(\frac{1}{2}\right)^{120/40}=\frac{1}{8}=0.125$.

43. The concentration of the drug decreases by a factor of $\left(\frac{1}{2}\right)^{24/16} \approx 0.354$ after 24 hours, and by a factor of $\left(\frac{1}{2}\right)^{72/16} = 0.0442$ after 72 hours.

45. In 30 years, the population will be $1{,}000{,}000 \times \left(\frac{1}{2}\right)^{30/24} \approx 420{,}000$, and it will be $1{,}000{,}000 \times \left(\frac{1}{2}\right)^{70/24} \approx 132{,}000$ in 70 years.

47. In 100 days, $1 \times \left(\frac{1}{2}\right)^{100/77} = 0.41$ kg of cobalt will remain, and $1 \times \left(\frac{1}{2}\right)^{200/77} = 0.17$ kg of cobalt will remain after 200 days.

49. $T_{\text{half}} \approx 70/6 = 11.7$ years; The fraction of the forest that will remain after 50 years is $(1-0.06)^{50} = 0.045$. Using the half-life formula, the fraction that will remain is $\left(\frac{1}{2}\right)^{50/11.7} \approx 0.052$, so the approximate half-life formula is not doing a reasonable job.

51. $T_{\text{half}} \approx 70/8 = 8.75$ years; In 2050 (34 years after 2016), the population will be $350{,}000(1-0.08)^{50} = 21{,}000$ elephants. Using the half-life formula, the population will be $350{,}000 \times \left(\frac{1}{2}\right)^{34/8.75} = 24{,}000$ elephants, so the approximate half-life formula does a somewhat reasonable job.

53. The approximate doubling time is $T_{\text{double}} \approx 70/6 = 11.7$ years. The exact doubling time is $\frac{\log_{10} 2}{\log_{10}(1+0.06)} \approx 11.9$ years. The price of a \$600 item in four years will be $\$600 \times 2^{4/11.9} = \$757.42$. (Note that this answer is using the rounded value for the "exact" doubling time – if you use all the digits your calculator stores, you should get a price of \$757.49).

55. The approximate doubling time is $T_{\text{double}} \approx 70/3.5 = 20$ years. The exact doubling time is $\frac{\log_{10} 2}{\log_{10}(1+0.035)} \approx 20.15$ years. The nation's population in 30 years will be $100{,}000{,}000 \times 2^{30/20.15} = 280{,}662{,}000$. (Note that this answer is using the rounded value for the "exact" doubling time – if you use all the digits your calculator stores, you should get a population of 280,679,370).

FURTHER APPLICATIONS

57. The amount of plutonium found today would be $10^{13} \times \left(\frac{1}{2}\right)^{4.6\times10^9/24{,}000} \approx 0$ tons. The reason there is no plutonium left is that 4.6 billion ÷ 24,000 is nearly 200,000, which means whatever amount of plutonium was originally present has undergone about 200,000 halving periods, and that's enough to reduce any amount imaginable to 0 (even if the entire mass of the earth were radioactive plutonium at the outset, it would have all decayed to other elements in 4.6 billion years).

59. $T_{\text{double}} \approx 70/4 = 17.5$ years; Using the doubling time formula, emissions would increase by a factor of $2^{35/17.5} = 4$. The exact formula for exponential growth estimates that emissions will increase by a factor of $(1.04)^{35} = 3.95$ from 2015 to 2050 (35 years).

61. The approximate half life is $T_{half} \approx 70/3 = 23.33$ years. The exact half life is $-\dfrac{\log_{10} 2}{\log_{10}(1-0.03)} \approx 22.76$ years. So we would expect the number of homicides to be reduced to 400 in about 23 years.

63. a. $1000 \text{ mb} \times \left(\dfrac{1}{2}\right)^{1/7} = 906 \text{ mb}$

b. $1000 \text{ mb} \times \left(\dfrac{1}{2}\right)^{8.848/7} = 416 \text{ mb}$

c. $\left(\dfrac{1}{2}\right)^{1/7} = 0.91$, so atmospheric pressure decreases by about 10% per kilometer.

TECHNOLOGY EXERCISES

69. a. $\log_{10} 500 \approx 2.698970$

b. $\log_{10} 25 \approx 1.397940$

c. $\log_{10}(1/125) \approx -2.096910$

d. $\log_{10} 0.2 \approx -0.698970$

e. $\log_{10} 0.05 \approx -1.301030$

f. $\log_{10} 625 \approx 2.795880$

UNIT 8C: REAL POPULATION GROWTH

QUICK QUIZ

1. **c.** Convert 80 million people per year into people per minute: $\dfrac{80\times10^6 \text{ people}}{\text{yr}} \times \dfrac{1 \text{ yr}}{365 \text{ d}} \times \dfrac{1 \text{ d}}{24 \text{ h}} \times \dfrac{1 \text{ hr}}{60 \text{ min}} = 152$ people per minute (or about 150 people/min).
2. **b.** The text states the world population doubled from 2 billion to 4 billion in approximately 52 years.
3. **b.** The birth rate peaked in 1960, and has seen a significant decline since then, but it is not the primary reason for the rapid growth – the decreasing death rate has played the most important role over the last three centuries.
4. **b.** This answer gives the correct definition of carrying capacity.
5. **c.** Answers **a** and **b** would tend to decrease the carrying capacity, while developing a cheap source of energy would allow us to solve some of the barriers to population growth, such as clean water for all.
6. **a.** The bacteria in a bottle exhaust their resources, at which point the population collapses.
7. **c.** The logistic model looks like exponential growth at the outset, but then the growth rate declines, and the population levels out to a level at (or below) the carrying capacity.
8. **a.** The birth rate in 1950 was quite high, and with no change in the death rate, this would lead to rapid exponential growth.
9. **a.** The approximate doubling time of a population growing at 1.1% per year is $T_{double} \approx 70/1.1 = 63.6$ years.
10. **c.** Answers **a** and **b** would be required in order for the population to level out (the birth rate must decline to equal the death rate if population is to remain steady, and a population of 10 billion is 50% larger than the current level of 7.1 billion, and thus we would need an increase of 50% in food production). This leaves answer **c**, which is not necessary to sustain a population of 10 billion.

REVIEW QUESTIONS

1. Before C.E. 1, the world population grew slowly. After that point, the population began to grow rapidly.
3. Birth rates have dropped rapidly throughout the world during the past 60 years, which is the same period that has seen the largest population growth in history. The rapid population growth comes from the fact that death rates have fallen even more dramatically than birth rates.

5. Logistic growth assumes that the population growth rate gradually decreases as the population approaches the carrying capacity. Given that the human population cannot grow exponentially forever, logistic growth means a sustainable future population. The alternative is overshoot and collapse, which might mean the end of our civilization.

DOES IT MAKE SENSE?

7. Makes sense. If the current growth rate of 1.1% remains steady for the next decade, we can expect a population of $7.5\times10^9\times(1.011)^{10}=8.4$ billion, which is an increase of 900 million people (i.e. more than twice the size of the current U.S. population of 321 million).

9. Does not make sense. The carrying capacity depends upon many factors, such as our ability to produce food for the current population.

11. Does not make sense. Predator/prey models often include the idea of overshoot and collapse.

BASIC SKILLS AND CONCEPTS

13. $T_{\text{double}}\approx70/0.9=78$ years; Under this assumption, the population in 2050 would be $7.1\times2^{37/78}\approx9.9$ billion.

15. $T_{\text{double}}\approx70/1.6=44$ years; Under this assumption, the population in 2050 would be $7.1\times2^{37/44}\approx12.7$ billion.

17. a. The growth rate in 1980 was $51.8-24.1=27.7$ per 1000 people. In 1995, it was $52.6-20.1=32.5$ per 1000 people. In 2016, it was $38.3-13.7=23.6$ per 1000 people.

b. The birth rate and death rate both declined overall, which resulted in a constant positive growth rate. This implies a growing population. If these trends continue, we can expect an increasing population in Afghanistan, though because there are so many factors that influence the growth rate of a country, it would be unwise to put much faith in these predictions 20 years hence.

19. a. The growth rate in 1980 was $16.0-11.3=4.7$ per 1000 people. In 1995, it was $10.9-14.6=-3.7$ per 1000 people. In 2016, it was $11.3-13.6=-2.3$ per 1000 people.

b. The growth rate fell to a negative value, which results in a decreasing population. If these trends continue, we can expect a slightly declining or stable population in Russia, though because there are so many factors that influence the growth rate of a country, it would be unwise to put much faith in these predictions 20 years hence.

21. When the population is 10 million, the actual growth rate is $4\%\times\left(1-\frac{10{,}000{,}000}{60{,}000{,}000}\right)\approx3.3\%$. When the population is 30 million, the actual growth rate is $4\%\times\left(1-\frac{30{,}000{,}000}{60{,}000{,}000}\right)\approx2.0\%$. When the population is 50 million, the actual growth rate is $4\%\times\left(1-\frac{50{,}000{,}000}{60{,}000{,}000}\right)\approx0.67\%$.

FURTHER APPLICATIONS

23. $T_{\text{double}}\approx70/0.7=100$ years; In 2050, the population will be $3.25\times10^8\times2^{33/100}=409$ million. In 2100, the population will be $3.25\times10^8\times2^{83/100}=578$ million.

25. $T_{\text{double}}\approx70/1.0=70$ years; In 2050, the population will be $3.25\times10^8\times2^{33/70}=451$ million. In 2100, the population will be $3.25\times10^8\times2^{83/70}=739$ million.

27. Answers will vary with student's age. $T_{\text{double}}\approx70/1.1=63.6$ years

For a student who was 18 in 2017: population at age 50 will be $7.5\times2^{(50-18)/63.6}=10.6$ billion; at age 80 the population will be $7.5\times2^{(80-18)/63.6}=14.7$ billion; at age 100 the population will be $7.5\times2^{(100-18)/63.6}=18.3$ billion

29. We must first compute the base growth rate r in the logistic model, using the 1960s data:

$$r = \frac{2.1\%}{\left(1 - \frac{3\text{ billion}}{9\text{ billion}}\right)} = 3.15\%.$$

The current growth rate can now be predicted using a population of 7.1 billion:

$$\text{growth rate} = 3.15\% \times \left(1 - \frac{7.5\text{ billion}}{9\text{ billion}}\right) = 0.525\%.$$

This is lower than the actual current growth rate of 1.1%, which indicates the assumed carrying capacity of 9 billion is too low to fit well with the data from the 1960s to present.

31. We must first compute the base growth rate r in the logistic model, using the 1960s data:

$$r = \frac{2.1\%}{\left(1 - \frac{3\text{ billion}}{15\text{ billion}}\right)} = 2.63\%.$$

The current growth rate can now be predicted using a population of 7.5 billion:

$$\text{growth rate} = 2.63\% \times \left(1 - \frac{7.5\text{ billion}}{15\text{ billion}}\right) = 1.31\%.$$

This is higher than the actual current growth rate of 1.1%, which indicates the assumed carrying capacity of 15 billion is too high to fit well with the data from the 1960s to present.

33. With an annual growth rate of 2%, the approximate doubling time would be $T_{\text{double}} \approx 70/2 = 35$ years. The population of the city in the year 2027 (ten years beyond 2017) would have been about $100{,}000(2)^{10/35} = 122{,}000$. The predicted populations in 2037 and 2077 would be 149,000 and 328,000, respectively (the computation is similar to the 2027 prediction). With an annual growth rate of 5%, the approximate doubling time would be $T_{\text{double}} \approx 70/5 = 14$ years. The population of the city in the year 2027 would have been $100{,}000(2)^{10/14} = 164{,}000$. The predicted populations in 2037 and 2077 would be 269,000 and 1,950,000, respectively.

UNIT 8D: LOGARITHMIC SCALES: EARTHQUAKES, SOUNDS, AND ACIDS

QUICK QUIZ

1. **a.** Each unit on the magnitude scale represents about 32 times as much energy as the previous magnitude.
2. **b.** Most deaths in earthquakes are caused by the collapse of buildings, landslides, and tsunamis.
3. **a.** By definition, a sound of 0 dB is the softest sound audible by a human.
4. **b.** From the definition of the decibel scale, we have $\frac{\text{intensity of sound}}{\text{intensity of softest audible sound}} = 10^{(95/10)} = 10^{9.5}$. The ratio on the left tells us this sound has intensity $10^{9.5}$ times as large as the softest audible sound.
5. **a.** From the definition of the decibel scale, we have $\frac{\text{intensity of sound}}{\text{intensity of softest audible sound}} = 10^{(10/10)} = 10$. The ratio on the left tells us a 10-dB sound is ten times as large as the softest audible sound, which is 0 dB.
6. **c.** Gravity follows the inverse square law, so its strength is proportional to the square of the distance between the objects. Thus if this distance is tripled (multiplied by 3), the strength of the force of gravity is decreased by a factor of $3^2 = 9$.
7. **c.** On the pH scale, a lower number corresponds to higher acidity.
8. **b.** Because $[H^+] = 10^{-\text{pH}}$, we have $[H^+] = 10^{-0} = 1$, so the hydrogen ion concentration is 1 mole per liter.

9. **c.** Because $pH = -\log_{10}[H^+]$, we have $pH = -\log_{10} 10^{-5} = -(-5) = 5$.

10. **a.** A lake damaged by acid rain has a relatively low pH. To bring it back to health, you want to raise the pH.

REVIEW QUESTIONS

1. The magnitude scale measures the strength of earthquakes. The magnitude is related to the energy released by the earthquake. Each magnitude represents about 32 times as much energy as the prior magnitude. For example, a magnitude 8 earthquake releases 32 times as much energy as a magnitude 7 earthquake.

3. pH measures the acidity of a substance using the concentration of positively charged hydrogen ions, which are hydrogen atoms without their electron. Acids have pH values lower than 7, bases have pH values higher than 7, and substances with a pH of 7 are neutral.

DOES IT MAKE SENSE?

5. Does not make sense. Each unit on the magnitude scale represents an increase of nearly 32 fold in the amount of energy released, and thus an earthquake of magnitude 6 releases about 32^3 (approximately 33,000) times as much energy as an earthquake of magnitude 3. This does not imply that a magnitude 6 earthquake will do 33,000 times as much damage, but it's not likely it will do only twice the damage of a magnitude 3 quake.

7. Does not make sense. The pH of water is related to its hydrogen ion concentration, which is not affected by its volume.

BASIC SKILLS AND CONCEPTS

9. $E = (2.5\times10^4)\times10^{1.5\times7} = 7.9\times10^{14}$ joules

11. $E = (2.5\times10^4)\times10^{1.5\times7.8} = 1.3\times10^{16}$ joules

13. The energy released by a magnitude 6 quake is 2.5×10^{13} joules (see Exercise 10), and since $\left(5\times10^{15}\right) \div \left(2.5\times10^{13}\right) = 200$, the bomb releases 200 times as much energy.

15. As shown in Table 8.5, a busy street is around 80 dB, and is $10^8 = 100$ million times as loud as the softest audible sound.

17. The loudness is $10\log_{10}\left(1\times10^6\right) = 60$ dB.

19. As with Example 4 in the text, we can compare the intensities of two sounds using the following: $\dfrac{\text{intensity of sound 1}}{\text{intensity of sound 2}} = 10^{(55-10)/10} = 31{,}623$. Thus a 55-dB sound is 31,623 times as loud as a 10-dB sound.

21. The distance has decreased by a factor of 4, which means the intensity of sound will increase by a factor of $4^2 = 16$.

23. The distance has decreased by a factor of 8, which means the intensity of sound will increase by a factor of $8^2 = 64$.

25. Each unit of increase on the pH scale corresponds to a decrease by a factor of 10 in the hydrogen ion concentration. Thus an increase of 5 on the pH scale corresponds to a decrease in the hydrogen ion concentration by a factor of $10^5 = 100{,}000$. This makes the solution more basic.

27. $[H^+] = 10^{-9.5} = 3.16\times10^{-10}$ mole per liter

29. $pH = -\log_{10}(0.0001) = 4$; This is an acid, as is any solution with pH less than 7.

31. Acid rain ($pH = 3$) is $10^3 = 1000$ times more acidic than ordinary water ($pH = 6$).

FURTHER APPLICATIONS

33. According to Table 8.4, earthquakes with magnitudes between 2 and 3 are labeled "very minor," and they occur about 1000 times per year worldwide. The glasses in the cupboards of the Los Angeles residents near the epicenter may rattle around a bit, but nothing serious would be expected to happen.

35. A solution with a pH of 12 is very basic: 100,000 times more basic than pure water. When children ingest such liquids, the mouth, throat, and digestive tracts are often burned. One would hope that a secondary effect of the incident would be that an adult would call the poison center immediately.

37. You probably would not be able to hear your friend (the noise of the sirens is louder). If your friend is shouting: "Watch out for that falling piano above you!" you might die.

39. a. The distance has increased by a factor of 100, so the intensity of the sound will decrease by a factor of $100^2 = 10{,}000$ times (this is due to the inverse square law). According to Table 8.5, busy street traffic is 80 dB, and is 10^8 times as loud as the softest audible sound. Note that a 40-dB sound is 10^4 times as loud as the softest sound, and thus is 10,000 times less intense than an 80-dB sound.

b. As with Example 4 in the text, we can compare the intensities of two sounds using the following:

$$\frac{\text{intensity of sound 1}}{\text{intensity of sound 2}} = 10^{(135-120)/10} = 31.6.$$

Thus a 135-dB sound is 31.6 times as loud as a 120-dB sound. In order to reduce the intensity of the sound by a factor of 31.6, you need to increase the distance from the speaker by a factor of $\sqrt{31.6} = 5.6$, so you should sit $5.6 \times 10 \text{ m} = 56$ meters away.

c. The distance from you to the booth (8 meters) is 8 times as much as the distance between the people talking (1 meter), so the intensity of the sound will have decreased by a factor of 64 by the time it reaches you. A 20-dB sound is 100 times louder than the softest audible sound, and when decreased by a factor of 64, it becomes (100/64) times as loud as the softest sound. Therefore its loudness in decibels is $10\log_{10}\left(\frac{100}{64}\right) = 1.94$ dB.

Note that a 60-dB sound is 10,000 times as loud as a 20-dB sound. If you wanted to amplify the sound that reaches you to 60 dB, you would need to increase it by a factor of 64 to overcome the loss of intensity due to distance (this would bring it back to 20 dB), and increase it again by a factor of 10,000 to raise it to the 60-dB level. This is equivalent to amplifying the sound by a factor of 640,000. Pray that the waiter doesn't come to your table and say a loud "hello" while your earpiece is in.

41. a. $[H^+] = 10^{-4}$ mole per liter.

b. The lake water with $\text{pH} = 7$ has a hydrogen ion concentration of 10^{-7} mole per liter. There are 3.785 liters in a gallon, so a 100 million gallon lake has $378{,}500{,}000 = 3.785 \times 10^8$ liters in it. Multiplying the hydrogen ion concentration by the volume of the lake gives us 37.85 moles of hydrogen ions before the acid was added. In a similar fashion, we find that 100,000 gallons (378,500 liters) of acid with $\text{pH} = 2$ $\left([H^+] = 10^{-2}\right)$ has 3785 moles of hydrogen ions. Combining both the volumes of the liquids and the moles of hydrogen ions gives us a hydrogen ion concentration of $\frac{37.85 + 3785 \text{ mole}}{3.785 \times 10^8 + 378{,}500 \text{ L}} = 10^{-5}$ mole per liter. This implies the polluted lake has $\text{pH} = 7 - \log_{10}\left(10^{-5}\right) = 5.$

c. Proceeding as in part (b), the acid-rain lake (without added acid) has 37,850 moles of hydrogen ions. When the acid is added (increasing the number of moles by 3785), the hydrogen ion concentration becomes $\frac{37{,}850 + 3785 \text{ mole}}{3.785 \times 10^8 + 378{,}500 \text{ L}} = 1.1 \times 10^{-4}$ mole per liter. This implies the polluted lake has $\text{pH} = 2 - \log_{10}(1.1 \times 10^{-4}) = 3.96.$

41. (continued)

d. The pollution could be detected if the acid was dumped into a lake with pH = 7 (part (b)), but it could not be detected if it was dumped into a lake with pH = 4 (part (c)).

UNIT 9A: FUNCTIONS: THE BUILDING BLOCKS OF MATHEMATICAL MODELS

QUICK QUIZ

1. **a.** A function describes how a dependent variable changes with respect to one or more independent variables.
2. **a.** The variable s is the independent variable, upon which the variable r depends.
3. **c.** The DJIA changes with respect to the passage of time, and thus is a function of time.
4. **b.** It is customary to plot the values of the independent variable along the horizontal axis.
5. **b.** The dependent variable is normally plotted along the vertical axis, and in this case, the dependent variable is z (with an independent variable of w).
6. **b.** The range of a function consists of the values of the dependent variable.
7. **a.** The speed of the car is the independent variable (because gas mileage changes with respect to speed), and the domain of a function is the set of values of interest for the independent variable.
8. **c.** The range of a function is the set of values corresponding to the values in the domain, which in turn are the set of values that make sense in the function. If you used a function to predict values outside the range, you would be forced to use values outside the domain; these values don't make sense.
9. **c.** Traffic volume on an urban freeway peaks in the morning rush hour, peaks again in the evening rush hour, and typically repeats this pattern day after day. If weekend traffic were significantly different from the weekday pattern, we would still see a weekly pattern repeating itself – this is a hallmark of a periodic function.
10. **b.** A model that does not agree well with past data probably won't do a good job of predicting the future.

REVIEW QUESTIONS

1. Mathematical models are a single equation or a set of equations used to represent real world situations, such as weather or traffic flow. If the assumptions of the model do not represent the complete situation, the predictions of the model may not be accurate. A model for daily temperature created only using morning temperatures will most likely not predict afternoon temperatures well.

3. Functions can be represented by a data table, a graph, or an equation or formula.

DOES IT MAKE SENSE?

5. Makes sense. Climatologists use mathematical models to predict and understand the nature of the earth's climate.

7. Does not make sense. If your heart rate depends upon your running speed, heart rate is the dependent variable, and the range is used to describe the values of the dependent variable.

BASIC SKILLS AND CONCEPTS

9.

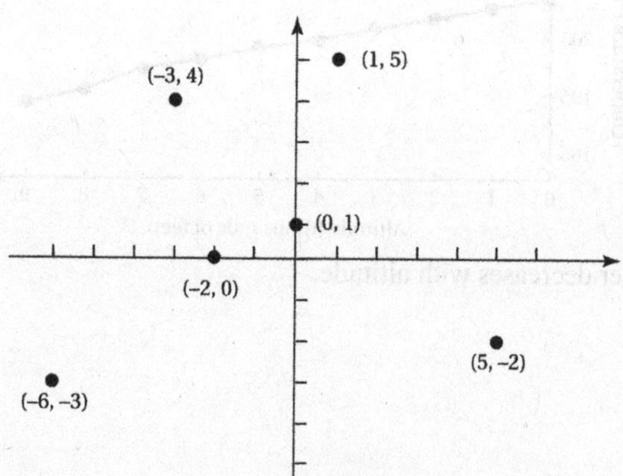

11. There is a functional relationship with *time* as the independent variable and *distance fallen* as the dependent variable.

13. There is a functional relationship with *price* as the independent variable and *demand* as the dependent variable.

15. As the volume of the tank increases, the cost of filling it also increases.

17. The temperature of the ocean decreases as latitude increases.

19. As the average speed of the car increases, the travel time decreases.

21. As the gas mileage increases, the cost of driving a fixed distance decreases (assuming that gas costs remain constant).

23. a. The pressure at 6000 feet is about 24 inches of mercury; the pressure at 18,000 feet is about 17 inches; and the pressure at 29,000 feet is about 11 inches.

b. The altitude is about 8000 feet when the pressure is 23 inches of mercury; the altitude is about 14,000 feet when the pressure is 19 inches; and the altitude is about 25,000 feet when the pressure is 13 inches.

c. It appears that the pressure reaches 5 inches of mercury at an altitude of about 50,000. The pressure approaches zero as one moves out of earth's atmosphere and into outer space.

25. a. On April 1, there's about 12.4 hours of daylight, and on October 31, about 10.7 hours. For all of these questions, your answers may vary somewhat from those given here, as it is difficult to read a graph of this sort with any level of precision.

b. There are 13 hours of daylight around April 19 and August 20.

c. There are 10.5 hours of daylight around January 31 and November 8.

d. At 20°N latitude (closer to the equator), the graph would be flatter (the variation in the length of a day is not as pronounced). At 60°N latitude (closer to the North Pole), the graph would be steeper and taller (the variation in the length of the day is larger). At 40°S latitude, the graph would have a similar shape, except that the location of the peaks and valleys would be exchanged (the shortest day of the year in the southern hemisphere is around June 21).

27. a. The independent variable is age (measured in years), and the dependent variable is weight (measured in pounds). The domain is the set of ages between 0 and 40 years, and the range is all weights between about 8 pounds and 130 pounds.

b. The function shows that weight increases with age, increasing rapidly in the early years and then leveling off in the later years.

29. a. The independent variable is altitude (in feet), and the dependent variable is boiling temperature (in °F). The domain is the set of altitudes between 0 and 9000 feet. The range is the set of temperatures between about 190°F and 212°F.

b.

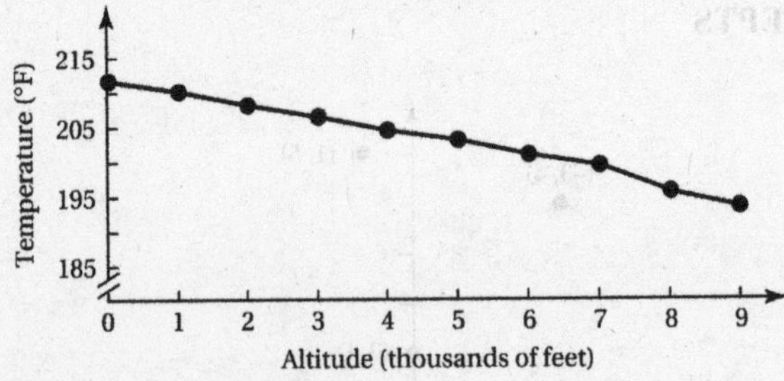

c. The boiling point of water decreases with altitude.

FURTHER APPLICATIONS

31. a. The domain of the function (*altitude, temperature*) is the set of altitudes of interest – say, 0 ft to 15,000 ft (or 0 m to 4000 m). The range is the set of temperatures associated with the altitudes in the domain; the interval 30°F to 90°F (or about 0°C to 30°C) would cover all temperatures of interest (this assumes a summer ascent of a mountain in the temperate zone of the earth).

b.

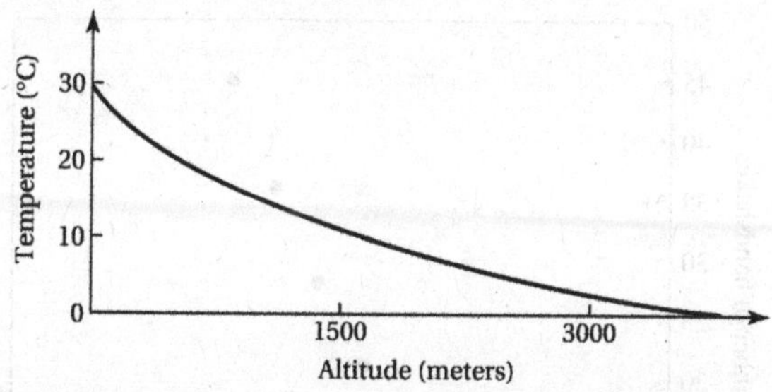

c. With some reliable data, this graph is a good model of how the temperature varies with altitude.

33. a. The domain of the function (*blood alcohol content, reflex time*) consists of all reasonable BACs (in gm/100 mL). For example, numbers between 0 and 0.25 would be appropriate. The range would consist of the reflex times associated with those BACs.

b.

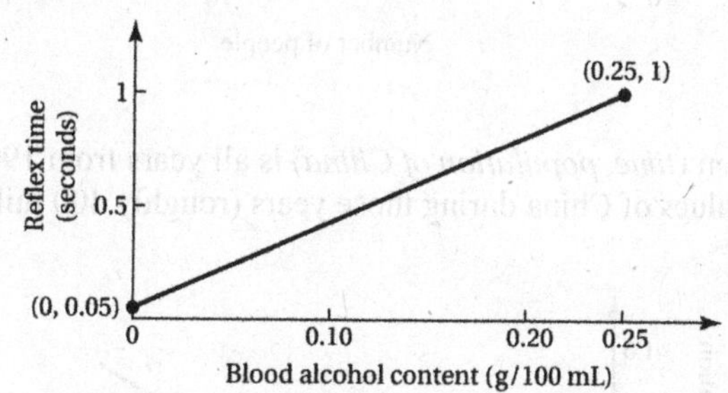

c. The validity of this graph as a model of alcohol impairment will depend on how accurately reflex times can be measured.

35. a. The domain of the function (*time of day, traffic flow*) consists of all times over a full day. The range consists of all traffic flows (in units of cars per minute) at the various times of the day. We would expect light traffic flow at night, medium traffic flow during the midday hours, and heavy traffic flow during the two rush hours.

b.

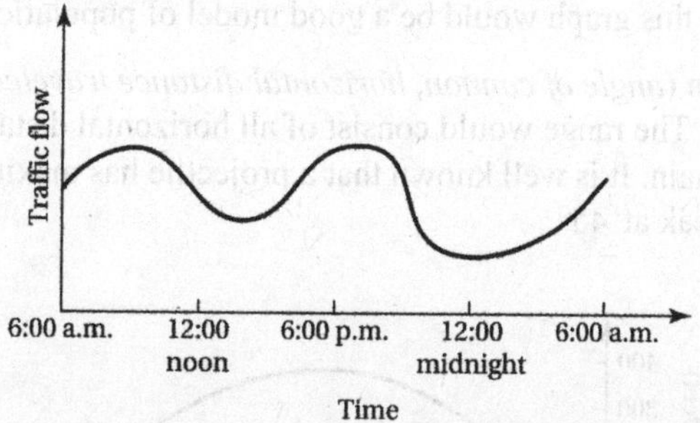

c. We would expect light traffic flow at night, medium traffic flow during the midday hours, and heavy traffic flow during the two rush hours. The graph of this function would be a good model only if based on reliable data.

37. a. The domain of the function (*number of people, number of handshakes*) consists of natural numbers from 2 (the minimum for a handshake to occur) to any arbitrary upper limit (for example, 10 people). The range consists of the number of handshakes, from 1 for two people up to the number for the upper limit of the domain. (Note: The formula for the number of handshakes between *n* people is $\frac{n(n-1)}{2}$.

b.

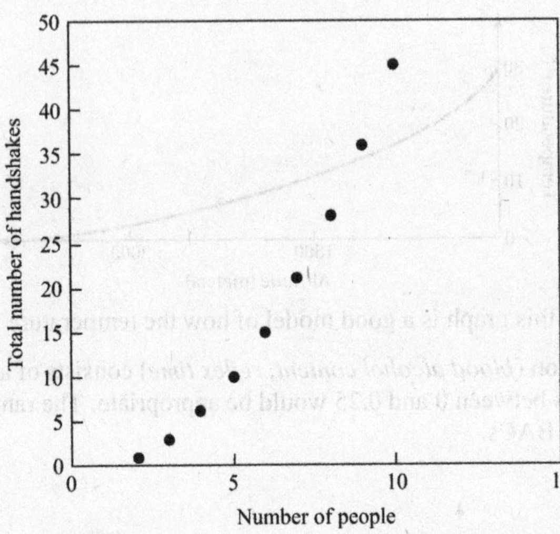

c. The model is exact.

39. a. The domain of the function (*time, population of China*) is all years from 1900 to, say, 2010. The range consists of the population values of China during those years (roughly 400 million to 1.3 billion).

b.

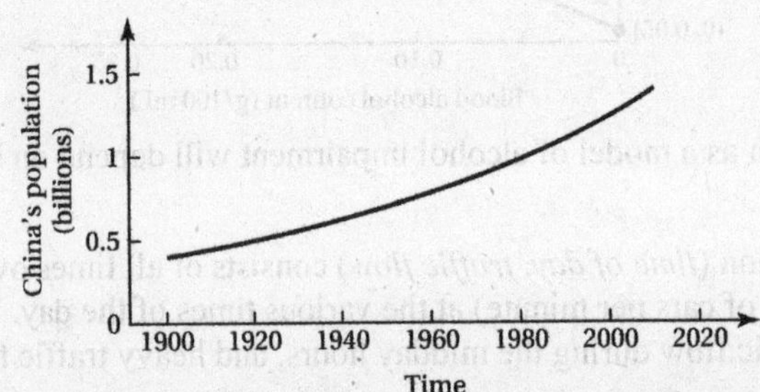

c. With accurate yearly data, this graph would be a good model of population growth in China.

41. a. The domain of the function (*angle of cannon, horizontal distance traveled by cannonball*) is all cannon angles between 0° and 90°. The range would consist of all horizontal distances traveled by the cannonball for the various angles in the domain. It is well known that a projectile has maximum range when the angle is about 45°, so the graph shows a peak at 45°.

b.

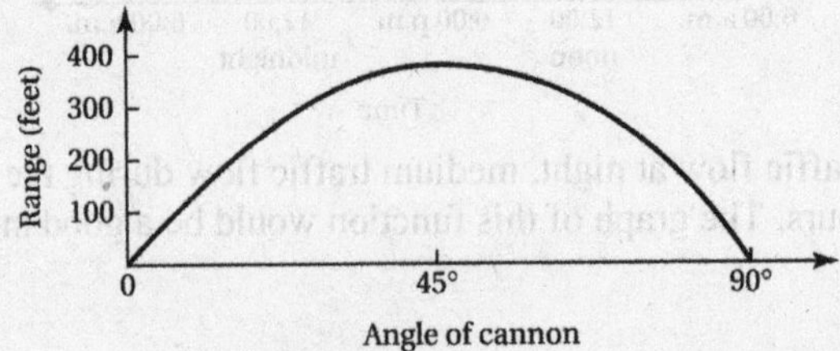

c. It is well known that a projectile has maximum range when the angle is about 45° so the graph above shows a peak at about 45°. This is a good qualitative model.

UNIT 9B: LINEAR MODELING

QUICK QUIZ

1. **c.** Constant absolute change over the same time interval (or other variable) gives rise to linear models, and this amounts to constant slope.
2. **b.** The slope of a linear function is the rate of change in the dependent variable with respect to the independent variable.
3. **c.** When a line slopes downward, the ratio of its change in y to change in x is negative, and this amounts to a negative slope or negative rate of change.
4. **c.** With only information about the first three miles of the trail, there's no knowing what will happen at mile 5.
5. **c.** Larger rates of change correspond with larger and steeper slopes.
6. **a.** The initial value corresponds with *time* = 0, and this results in an initial *price* = $100.
7. **c.** The coefficient on the independent variable is the slope, which also represents the rate of change. Since this coefficient is negative, we can expect a graph that slopes downward.
8. **a.** The y-intercept is $b = 7$ and the slope is $m = -2$, and thus the line, in the form $y = mx + b$, has equation $y = -2x + 7$.
9. **b.** A line in the form $y = mx + b$ has slope of m and y-intercept of b, so the slope of both $y = 12x - 3$ and $y = 12x + 3$ is $m = 12$, and they have different y-intercepts (-3 and 3).
10. **c.** Charlie picks $550 - 150 = 400$ apples over the span of two hours, so the slope of the function is $400/2 = 200$.

REVIEW QUESTIONS

1. A function is linear if it has a constant rate of change. A linear function will have a straight-line graph.
3. The rate of change of a linear function is equal to the slope of the graph of the function.
5. The general formula is *dependent variable* = initial value + (rate of change × *independent variable*), which is equivalent to $y = mx + b$, the standard algebraic form.

DOES IT MAKE SENSE?

7. Does not make sense. The graph of a linear function is always a line.
9. Makes sense. The familiar relationship *distance* = *speed* × *time* can be solved for *speed* to produce *speed* = *distance* ÷ *time*, which is the rate of change in distance with respect to time.

BASIC SKILLS AND CONCEPTS

11. a. Rain depth increases linearly with time.

    b. It appears that the change in depth over the first three hours is four inches, so the slope is $4/3$ inches per hour.

    c. The model is realistic if the rainfall rate is a constant $4/3$ inches per hour over four hours.

13. a. On a long trip, the distance from home decreases linearly with time.

    b. The distance has decreased by 500 miles over the span of 7 hours, so the slope is $\frac{-500 \text{ mi}}{7 \text{ hr}} = -71.4$ miles per hour.

    c. This is a good model if the speed of travel is a constant 71.4 miles per hour over 7 hours.

15. a. Shoe size increases linearly with the height of the individual.

b. The change in shoe size was 11 while the change in height was 80 inches, and this produces a slope of $\frac{11 \text{ size}}{80 \text{ in}} = 0.1375$ size per inch.

c. This model is a rough approximation at best, and it would be difficult to find realistic conditions where the relationship held.

17. The *water depth* decreases with respect to *time* at a rate of 2 in/day, so the rate of change is -2 in/day. In 8 days, the water depth decreases by $2 \text{ in/day} \times 8 \text{ days} = 16$ inches. In 15 days, it decreases by $2\text{in/day} \times 15 \text{ day} = 30$ inches.

19. The *Fahrenheit temperature* increases with respect to the *Celsius temperature* at a rate of 9 °F per 5 °C. The rate of change is $\frac{9}{5}$ °F/°C. An increase of 5°C results in an increase of $5°\text{C} \times \frac{9}{5}°\text{F/°C} = 9°\text{F}$. A decrease of 25°C results in a decrease of $25°\text{C} \times \frac{9}{5}°\text{F/°C} = 45°\text{F}$.

21. Tony's *weight* decreases with respect to *time* at a rate of 0.6 pounds per day. After 15 days, his weight will have decreased by $0.6 \text{ lb/day} \times 15 \text{ days} = 9$ pounds. After 4 weeks (28 days), his weight will have decreased by $0.6 \text{ lb/day} \times 28 \text{ days} = 16.8$ pounds.

23. The independent variable is time t, measured in years, where $t = 0$ represents today. The dependent variable is price p, measured in dollars. The equation for the price function is $p = 18{,}000 + 900t$. In 3.5 years, a new car will cost $p = 18{,}000 + 900 \times 3.5 = \$21{,}150$. This function does not give a good model for car prices.

25. The independent variable is snow depth d, measured in inches, and the dependent variable is maximum speed s, measured in miles per hour. The equation for the speed function is $s = 40 - 1.1d$. To find the depth at which the plow will not be able to move, set $s = 0$ and solve for d.

$$0 = 40 - 1.1d \Rightarrow 1.1d = 40 \Rightarrow d = 40/1.1 = 36 \text{ inches.}$$

The rate at which the speed decreases per inch of snow depth is probably not a constant, so this model is an approximation.

27. The independent variable is time t, measured in minutes, and the dependent variable is rental cost r, measured in dollars. As long as the copy business is willing to prorate rental charges per minute (rather than in 5-minute blocks, even if you do not use the full 5 minutes), the change per minute is $\$2.00/5 = \$0.40$ per minute. This implies the rental cost function is $r = 10 + 0.40t$. To find out how many minutes can be rented for \$25, set $r = 25$, and solve for t.

$$25 = 10 + 0.40t \Rightarrow 15 = 0.40t \Rightarrow t = 15/0.40 = 37.5 \text{ minutes.}$$

This function gives a very good model of rental costs, provided all of the costs are quoted correctly

29. Let W represent the weight of the dog in pounds, and t the time in years. The information given about the weight of your dog correspond to the points $(0, 5.5)$ and $(1, 20)$. The slope of the model is $\frac{20 - 5.5}{1} = 14.5$ pounds per year, so the linear function is $W = 14.5t + 5.5$. When the dog is 5 years old, its weight is $W = 14.5 \times 5 + 5.5 = 78$ pounds, and at age 10, it weighs $W = 14.5 \times 10 + 5.5 = 150.5$ pounds. The model is accurate only for lower ages.

31. Let P represent the profit (or loss) realized when selling n raffle tickets. The initial cost of the raffle to the fundraisers is \$350 (a negative profit), and the rate of change of the profit with respect to the number of tickets sold is \$10 per ticket. Thus the profit function is $P = 10n - 350$. Set $P = 0$ and solve for n to find out how many tickets must be sold for the club to break even, which is 35.

$$0 = 10n - 350 \Rightarrow 10n = 350 \Rightarrow n = 350/10 = 35 \text{ tickets}$$

33. Let V represent the value of the washing machine, and let t represent time (in years). The value function is $V = -125t + 1500$. Set $V = 0$ and solve for t to find out how long it will take for the value to reach \$0, which is 12 years.

$$0 = -125t + 1500 \Rightarrow 125t = 1500 \Rightarrow t = 1500/125 = 12 \text{ years}$$

FURTHER APPLICATIONS

35. The slope is $m = 2$. The y-intercept is $(0, 6)$.

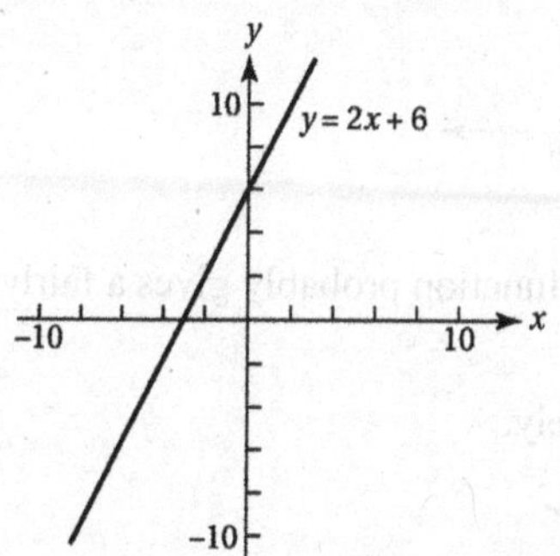

37. The slope is $m = -5$; The y-intercept is $(0, -5)$.

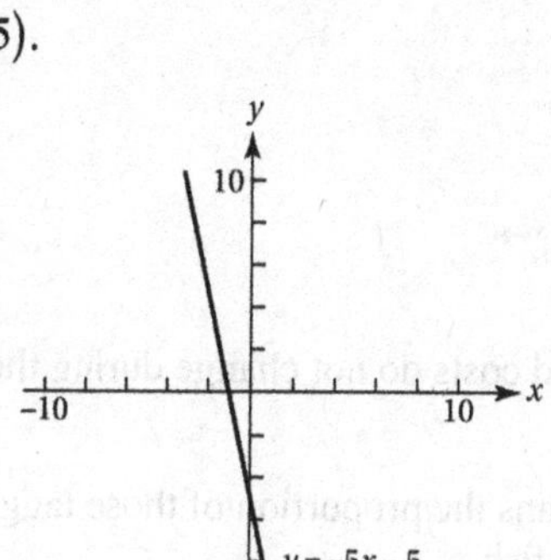

39. The slope is $m = 3$. The y-intercept is $(0, -6)$.

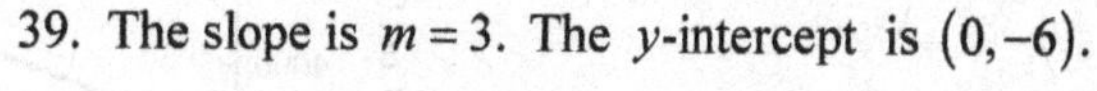

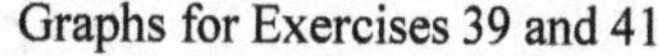
Graphs for Exercises 39 and 41

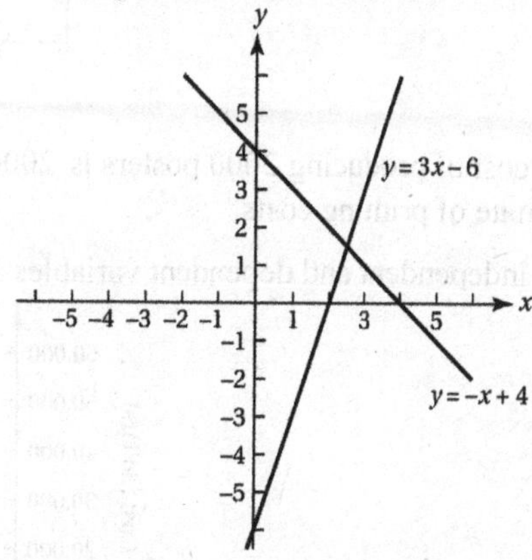

41. The slope is $m = -1$. The y-intercept is $(0, 4)$.

43. The independent and dependent variables are *time* and *elevation*, respectively.

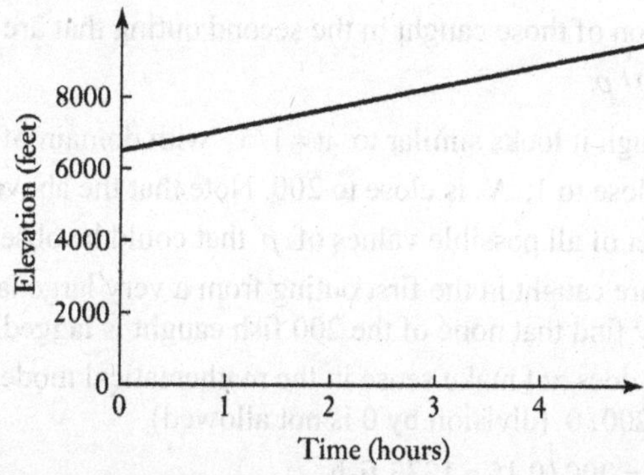

After 3.5 hours, the elevation of the climbers is $6500 + 600 \times 3.5 = 8600$ feet. The model is reasonable provided the rate of ascent is nearly constant.

45. The independent and dependent variables are *number of posters* and *cost*, respectively.

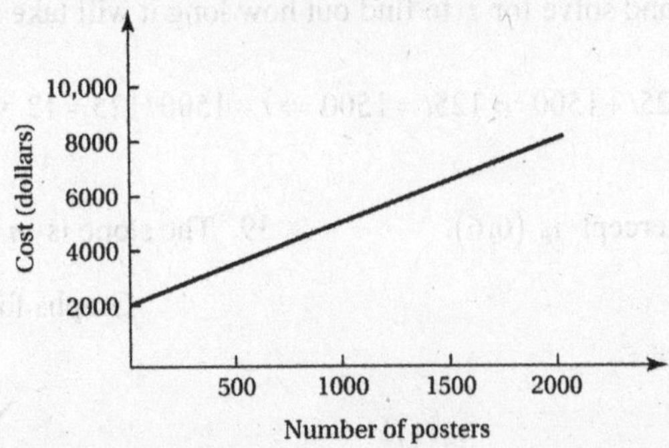

The cost of producing 2000 posters is $2000 + 2000 \times 3 = \$8000$. This function probably gives a fairly realistic estimate of printing costs.

47. The independent and dependent variables are *time* and *cost*, respectively.

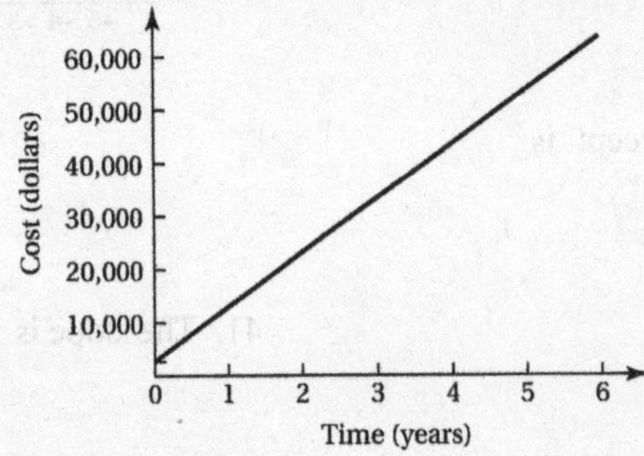

The cost of 6 years of school is $2000 + 10{,}000 \times 6 = \$62{,}000$. Provided costs do not change during the six-year period, this function is an accurate model of the cost.

49. a. Half of the fish caught in the second outing are tagged, and this means the proportion of those tagged in the entire population is 1/2. Thus $200 / N = 1/2$, which implies $N = 400$ fish.

b. One-fourth of the fish caught in the second outing are tagged, and this means the proportion of those tagged in the entire population is 1/4. Thus $200 / N = 1/4$, which implies $N = 800$ fish.

c. If p represents the proportion of those caught in the second outing that are tagged, we have $200 / N = p$, which implies $N = f(p) = 200 / p$.

d. The graph not supplied, though it looks similar to $y = 1/x$, with domain of $0 < p \le 1$. When p is small, N is very large, and when p is close to 1, N is close to 200. Note that the above domain is the domain of the mathematical model, not the set of all possible values of p that could be observed in such a study. Consider a situation where only 200 fish are caught in the first outing from a very large lake with a large fish population. On the second outing, you may find that none of the 200 fish caught is tagged, in which case $p = 0$. This makes sense in the field, but it does not make sense in the mathematical model, because we cannot predict the fish population by using $N = 200/0$ (division by 0 is not allowed).

e. If $p = 15\% = 0.15$, then $N = 200/0.15 \approx 1333$ fish.

f. Answers will vary.

UNIT 9C: EXPONENTIAL MODELING

QUICK QUIZ

1. **b.** This is the definition of exponential growth.
2. **a.** Q_0 represents the initial value in the exponential model.
3. **b.** The variable r represents the growth rate, expressed as a decimal (and $3\% = 0.03$).

4. **a.** The population of India at any time t is given by $1,340,000,000 \times 1.012^t$, where t is years since 2017.

5. **c.** Because the value of the dollar is decreasing, r is negative.

6. **c.** If you start out with a positive quantity, and remove half of it, half still remains (and it is a positive quantity, which can never equal 0).

7. **b.** If you know the initial amount and half-life of any exponentially decreasing quantity, you can use

   $Q = Q_0 \times \left(\frac{1}{2}\right)^{t/T_{\text{half}}}$ to predict the future amount after any time t.

8. **c.** You need to know both the initial amount Q_0 and the amount at some future time t in order to solve the exponential model $Q = Q_0 \times \left(\frac{1}{2}\right)^{t/T_{\text{half}}}$ for t (you must also know T_{half}, but we are given that).

9. **c.** Because 1/8 of the original uranium is present, we know it has undergone 3 halving periods, and since each of these is 700 million years, the rock is 3×700 million = 2.1 billion years old.

10. **a.** We know $Q = Q_0(1+r)^t$, and $Q = Q_0 \times 2^{t/T_{\text{double}}}$, and thus $Q_0(1+r)^t = Q_0 \times 2^{t/T_{\text{double}}}$. Divide both sides by Q_0 to get the result shown in the answer.

REVIEW QUESTIONS

1. Q = the value of the exponentially growing (or decaying) quantity at time t; Q_0 = the initial value of the quantity (at $t = 0$); r = the fractional growth (or decay) rate for the quantity; t = time; The units of time used for t and r must be the same.

3. Start at the point $(0, Q_0)$ that represents the initial value at $t = 0$. For an exponentially growing quantity, we know that the value of Q is $2Q_0$ (double its initial value) after one doubling time (T_{double}), $4Q_0$ after two doubling times ($2T_{\text{double}}$), $8Q_0$ after three doubling times ($3T_{\text{double}}$), and so on. We simply fit a steeply rising curve between these points. Using half life, follow the same procedure using $\frac{1}{2}Q_0$, $\frac{1}{4}Q_0$, and $\frac{1}{8}Q_0$ for T_{half}, $2T_{\text{half}}$, and $3T_{\text{half}}$, respectively. See Figure 9.18 on page 562 for the general shape of the graphs of exponential growth and decay.

5. Exponential functions can be used to determine the change in prices over time due to inflation, predict future populations and energy use, calculate the levels of medicine in the bloodstream, and determine the age of fossils.

DOES IT MAKE SENSE?

7. Does not make sense. After 100 years, the population growing at 2% per year will have increased by a factor of $1.02^{100} = 7.245$, whereas the population growing at 1% per year will have increased by a factor of $1.01^{100} = 2.705$. Under only very special circumstances will the first population grow by twice as many people as the second population. (Let Q_0 represent the initial amount of the population growing by 2%, and R_0 the initial amount of the population growing by 1%. After 100 years, Q_0 will have grown to $7.245Q_0$, and thus will have grown by $7.245Q_0 - Q_0 = 6.245Q_0$. In a similar fashion, it can be shown that R_0 will have grown by $1.705R_0$. We need $6.245Q_0 = 2 \times 1.705R_0$ in order to satisfy the condition that the first population will grow by twice as many people as the second population. This implies that the ratio $R_0/Q_0 = 6.245/3.41 = 1.83$. In other words, whenever the initial amount of the second population is 1.83 times as large as the initial amount of the first population, the first population will grow by twice as many people in 100 years. (Try it with initial populations of 100 and 183.)

9. Makes sense. If we know the half-life of the radioactive material, we can create an exponential function that models the quantity remaining as time passes. In order to use this model to determine ages of bones, we only need to measure the amount of radioactive material present in the bone, and the amount it had originally (the latter is usually surmised by making assumptions about conditions present at the time of death).

BASIC CONCEPTS AND SKILLS

11. $2^x = 128 \Rightarrow \log_{10} 2^x = \log_{10} 128 \Rightarrow x\log_{10} 2 = \log_{10} 128 \Rightarrow x = \dfrac{\log_{10} 128}{\log_{10} 2} = 7$

13. $3^x = 99 \Rightarrow \log_{10} 3^x = \log_{10} 99 \Rightarrow x\log_{10} 3 = \log_{10} 99 \Rightarrow x = \dfrac{\log_{10} 99}{\log_{10} 3} = 4.18$

15. $7^{3x} = 623 \Rightarrow \log_{10} 7^{3x} = \log_{10} 623 \Rightarrow 3x\log_{10} 7 = \log_{10} 623 \Rightarrow x = \dfrac{\log_{10} 623}{3\log_{10} 7} = 1.10$

17. $9^x = 1748 \Rightarrow x = \dfrac{\log_{10} 1748}{\log_{10} 9} = 3.40$ (See Exercise 13 for a similar process of solution.)

19. $\log_{10} x = 4 \Rightarrow 10^{\log_{10} x} = 10^4 \Rightarrow x = 10{,}000$

21. $\log_{10} x = 3.5 \Rightarrow 10^{\log_{10} x} = 10^{3.5} \Rightarrow x = 10^{3.5} = 3162.28$

23. $3\log_{10} x = 4.2 \Rightarrow \log_{10} x = 1.4 \Rightarrow x = 10^{1.4} = 25.12$

25. $\log_{10}(4+x) = 1.1 \Rightarrow 4 + x = 10^{1.1} \Rightarrow x = 8.59$

27. a. $Q = 60{,}000 \times (1.025)^t$, where Q is the population of the town, and t is time (measured in years).

b.

Year	Population
0	60,000
1	61,500
2	63,038
3	64,613
4	66,229
5	67,884
6	69,582
7	71,321
8	73,104
9	74,932
10	76,805

c.

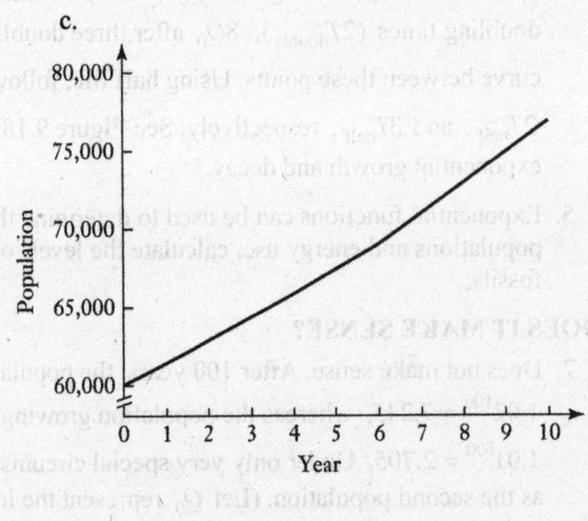

29. a. $Q = 1{,}000{,}000 \times (0.93)^t$, where Q is the number of acres of forest left after t years.

b.

Year	Millions of Acres
0	1.00
1	0.93
2	0.86
3	0.80
4	0.75
5	0.70
6	0.65
7	0.60
8	0.56
9	0.52
10	0.48

c.

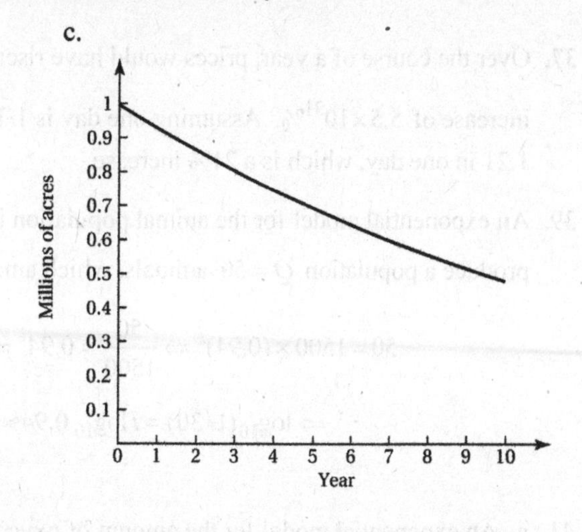

31. a. $Q = 175{,}000 \times (1.05)^t$, where Q is the average price of a home in dollars, and t is time, measured in years, with $t = 0$ corresponding to 2013.

b.

Year	Average price
0	\$175,000
1	\$183,750
2	\$192,938
3	\$202,584
4	\$212,714
5	\$223,349
6	\$234,517
7	\$246,243
8	\$258,555
9	\$271,482
10	\$285,057

c.

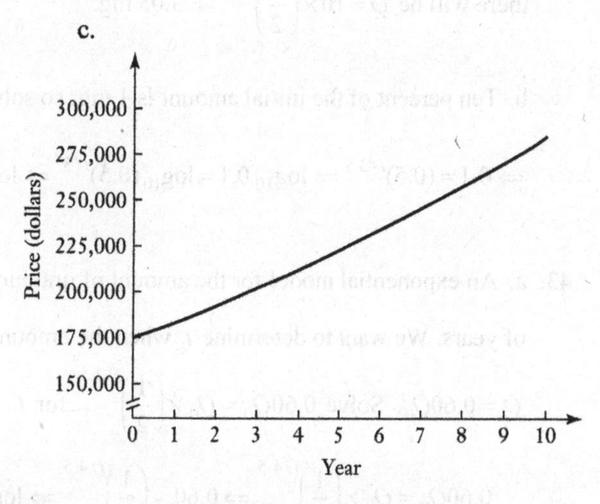

33. a. $Q = 2000 \times (1.05)^t$, where Q is your monthly salary in dollars, and t is time, measured in years.

b.

Year	Monthly Salary
0	\$2000.00
1	\$2100.00
2	\$2205.00
3	\$2315.25
4	\$2431.01
5	\$2552.56
6	\$2680.19
7	\$2814.20
8	\$2954.91
9	\$3102.66
10	\$3257.79

c.

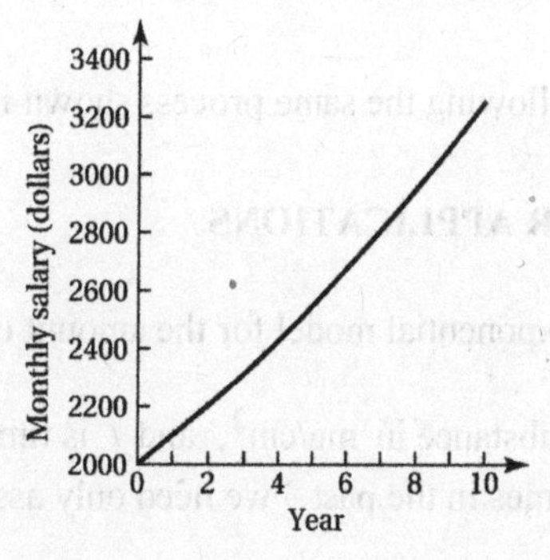

35. Over the span of a year, prices will increase by a factor of $(1.025)^{12} = 1.345$, which is a 34.5% increase.

37. Over the course of a year, prices would have risen by a factor of $\left(1+\frac{30,000}{100}\right)^{12} = 5.5\times10^{29}$, or an annual increase of 5.5×10^{31}%. Assuming one day is 1/30 of a month, prices would rise by a factor of $(301)^{(1/30)} =$ 1.21 in one day, which is a 21% increase.

39. An exponential model for the animal population is $Q = 1500\times(0.94)^t$. We need to find the time t that will produce a population $Q = 50$ animals, which amounts to solving $50 = 1500\times(0.94)^t$ for t.

$$50 = 1500\times(0.94)^t \Rightarrow \frac{50}{1500} = 0.94^t \Rightarrow (1/30) = 0.94^t \Rightarrow \log_{10}(1/30) = \log_{10} 0.94^t$$

$$\Rightarrow \log_{10}(1/30) = t\log_{10} 0.94 \Rightarrow t = \frac{\log_{10}(1/30)}{\log_{10} 0.94} = 54.97, \text{ or 55 years}$$

41. a. An exponential model for the amount of oxycodone in the bloodstream is $Q = 10\times\left(\frac{1}{2}\right)^{t/3.5}$. After 6 hours, there will be $Q = 10\times\left(\frac{1}{2}\right)^{6/3.5} = 3.05$ mg.

b. Ten percent of the initial amount is 1 mg, so solve $1 = 10\times\left(\frac{1}{2}\right)^{t/3.5}$ for t: $1 = 10\times\left(\frac{1}{2}\right)^{t/3.5}$

$$\Rightarrow 0.1 = (0.5)^{t/3.5} \Rightarrow \log_{10} 0.1 = \log_{10}(0.5)^{t/3.5} \Rightarrow \log_{10} 0.1 = \frac{t}{3.5}\log_{10} 0.5 \Rightarrow t = \frac{3.5\log_{10} 0.1}{\log_{10} 0.5} \Rightarrow t = 11.6 \text{ hours}$$

43. a. An exponential model for the amount of uranium left is $Q = Q_0\times\left(\frac{1}{2}\right)^{t/4.5}$, where t is measured in billions of years. We want to determine t when the amount Q is equal to 60% of the original amount, i.e. when $Q = 0.60Q_0$. Solve $0.60Q_0 = Q_0\times\left(\frac{1}{2}\right)^{t/4.5}$ for t. Begin by dividing both sides by Q_0.

$$0.60Q_0 = Q_0\times\left(\frac{1}{2}\right)^{t/4.5} \Rightarrow 0.60 = \left(\frac{1}{2}\right)^{t/4.5} \Rightarrow \log_{10} 0.60 = \log_{10}(0.5)^{t/4.5} \Rightarrow \log_{10} 0.60 = \frac{t}{4.5}\log_{10} 0.5$$

$$\Rightarrow t = \frac{4.5\log_{10} 0.60}{\log_{10} 0.5} \Rightarrow t \approx 3.3 \text{ billion years}$$

b. Following the same process shown in part (a), we have $t = \frac{4.5\log_{10} 0.55}{\log_{10} 0.5} = 3.9$ billion years old.

FURTHER APPLICATIONS

45. An exponential model for the amount of radioactive substance is $Q = 3\times\left(\frac{1}{2}\right)^{t/20}$, where Q is the density of the substance in mg/cm^2, and t is time in years, with $t = 0$ corresponding to the present. This model is valid for times in the past – we need only assign negative values for t. Thus 55 years ago corresponds with $t = -55$, so the density at that time was $Q = 3\times\left(\frac{1}{2}\right)^{-55/20} = 20.2$ mg/cm^2. One can also solve $3 = Q_0\times\left(\frac{1}{2}\right)^{55/20}$ for Q_0 to produce the same result.

47. The median home price would be $Q = \$300,000 \times (1.11)^{17} = \$1,768,527$, which is much higher that the actual median price of \$400,000.

49. a. Experiment with various values of r in $407 = 316 \times (1+r)^{58}$ to find that $r \approx 0.0044$ or 0.44% per year. One can also solve the equation using algebra: $407 = 316 \times (1+r)^{58} \Rightarrow \frac{407}{316} = (1+r)^{58}$. Now raise both sides to the reciprocal power of 58, i.e. to the 1/58 power, and then subtract 1 to find $r = \left(\frac{407}{316}\right)^{1/58} - 1$ $= 0.0044$ or 0.44%.

b. We want to determine t when $Q = 560$. Solve $560 = 316 \times (1.0044)^t$ for t. The CO_2 concentration will be 560 level in $1959 + 130 = 2089$.

$$560 = 316 \times (1.0044)^t \Rightarrow \frac{560}{316} = (1.0044)^t \Rightarrow \log_{10}(140/79) = \log_{10}\left(1.0044^t\right)$$

$$\Rightarrow \log_{10}(140/79) = t \times \log_{10} 1.0044 \Rightarrow t = \frac{\log_{10}(140/79)}{\log_{10} 1.0044} \Rightarrow t \approx 130.33 \text{ years}$$

c. Answers will vary.

UNIT 10A: FUNDAMENTALS OF GEOMETRY

QUICK QUIZ

1. **a.** As long as the points are distinct, two points are sufficient to determine a unique line.
2. **b.** The surface of a wall is similar to a plane, which is two-dimensional.
3. **c.** By definition, an obtuse angle has measure more than $90°$.
4. **c.** By definition, a regular polygon is a many-sided figure where each side has the same length, and each internal angle is the same.
5. **b.** By definition, a right triangle has one $90°$ angle.
6. **a.** The formula for the circumference of a circle of radius r is $C = 2\pi r$.
7. **c.** The formula for the volume of a sphere of radius r is $V = \frac{4}{3}\pi r^3$.
8. **c.** Suppose the side length of a square is denoted by s. Its area is then s^2. If the lengths of the sides are doubled to $2s$, the area becomes $(2s)^2 = 4s^2$, and this is 4 times as large as the original area.
9. **c.** The volume of a sphere whose radius has tripled from r to $3r$ is $V = \frac{4}{3}\pi(3r)^3 = 3^3 \cdot \frac{4}{3}\pi r^3$.
10. **a.** The volume of a block does not increase when it is cut into pieces, though its surface area does, which implies is surface-area-to-volume ratio does as well.

REVIEW QUESTIONS

1. Euclidean geometry is the geometry described in Euclid's work. It is the familiar geometry of lines, angles, and planes.
3. Dimensions are the number of independent directions in which you could move if you were on an object. The number of coordinates needed to locate a point is the same as the dimension of the object on which the point lies.
5. Plane geometry is the geometry of two-dimensional objects. Two lines or line segments that meet in a right angle ($90°$) are said to be perpendicular. Two lines or line segments in a plane that are the same distance apart (never meet) everywhere along their lengths are said to be parallel. (See Figure 10.5 on page 581.)
7. circumference of circle $= 2 \times \pi \times$ radius; area of circle $= \pi \times \text{radius}^2$
9. Lengths scale with the scale factor, areas scale with the square of the scale factor, and volumes scale with the cube of the scale factor. For example, a scaling factor of 3 will increase lengths by a factor of $3^1 = 3$, areas by a factor of $3^2 = 9$, and volumes by a factor of $3^3 = 27$.

DOES IT MAKE SENSE?

11. Does not make sense. If the highways are, in fact, straight lines, they will not intersect more than once.
13. Makes sense. A rectangular prism is just a box.
15. Does not make sense. Basketballs are in the shape of a sphere (a can is shaped like a right circular cylinder).

BASIC SKILLS AND CONCEPTS

17. One-half of a circle is $180°$.
19. One-fifth of a circle is $72°$.
21. Two-ninths of a circle is $80°$.
23. A $45°$ angle subtends $45°/360° = 1/8$ of a circle.
25. A $120°$ angle subtends $120°/360° = 1/3$ of a circle.
27. A $36°$ angle subtends $36°/360° = 1/10$ of a circle.

29. A $300°$ angle subtends $300°/360° = 5/6$ of a circle.

31. $C = 2\times\pi\times(6\text{ m}) = 12\pi\text{ m} \approx 37.7\text{ m}$

$A = \pi\times(6\text{ m})^2 = 36\pi\text{ m}^2 \approx 113.1\text{ m}^2$

33. $C = 2\times\pi\times\left(\frac{23\text{ ft}}{2}\right) = 23\pi\text{ ft} \approx 72.3\text{ ft}$

$A = \pi\times\left(\frac{23\text{ ft}}{2}\right)^2 = \frac{529\pi}{4}\text{ ft}^2 \approx 415.5\text{ ft}^2$

35. $C = 2\times\pi\times\left(\frac{40\text{ mm}}{2}\right) = 40\pi\text{ mm} \approx 125.7\text{ mm}$

$A = \pi\times\left(\frac{40\text{ mm}}{2}\right)^2 = 400\pi\text{ mm}^2 \approx 1256.6\text{ mm}^2$

37. The perimeter is $4\times 8\text{ mi} = 32\text{ mi}$. The area is $8\text{ mi}\times 8\text{ mi} = 64\text{ mi}^2$.

39. The perimeter is $2\times 8\text{ ft} + 2\times 30\text{ ft} = 76\text{ ft}$. The area is $4\text{ ft}\times 30\text{ ft} = 120\text{ ft}^2$.

41. The perimeter is $2\times 2.2\text{ cm} + 2\times 2.0\text{ cm} = 8.4\text{ cm}$. The area is $2.2\text{ cm}\times 2.0\text{ cm} = 4.4\text{ cm}^2$.

43. The perimeter is $6+8+10=24$ units. The area is $\frac{1}{2}\times 8\times 6 = 24\text{ units}^2$.

45. The perimeter is $8+5+5=18$ units. The area is $\frac{1}{2}\times 8\times 3 = 12\text{ units}^2$.

47. The two semicircular caps together make a single circle, whose radius is 3 feet. The perimeter, then, is $2\times 8\text{ ft} + 2\pi\times 3\text{ ft} = 34.8\text{ ft}$ and the area is $6\times 8\text{ ft} + \pi\times(3\text{ ft})^2 = 76.27\text{ ft}^2$.

49. The area is $\frac{1}{2}\times 12\text{ ft}\times 11\text{ ft} = 66\text{ ft}^2$.

51. The area of the parallelogram is $180\text{ yd}\times 150\text{ yd} = 27{,}000\text{ yd}^2$

53. The pool holds $50\text{ m}\times 30\text{ m}\times 2.5\text{ m} = 3750\text{ m}^3$.

55. The duct is in the shape of a cylinder, and so its volume is $\pi\times(1.5\text{ ft})^2\times 40\text{ ft} = 282.7\text{ ft}^3$ (note that 18 inches is 1.5 feet). The surface area of the duct is $2\pi\times 1.5\text{ ft}\times 40\text{ ft} = 377.0\text{ ft}^2$. We are assuming here that there are no circular end pieces on the duct that need to be painted. If there are end pieces, the area would increase by $2\pi\times(1.5\text{ ft})^2 = 14.1\text{ ft}^2$.

57. The circumference is greater because it is π diameters (of a tennis ball) long, whereas the height is only 3 diameters tall.

59. The volume of the reservoir is $250\text{ m}\times 60\text{ m}\times 12\text{ m} = 180{,}000\text{ m}^3$. 30% of this volume, which is $54{,}000\text{ m}^3$, must be added.

61. The first tree has greater volume ($\pi\times 2.5^2\times 40 = 785.4\text{ ft}^3$) than the second tree (whose volume is $\pi\times 2.1^2\times 50 = 692.7\text{ ft}^3$).

63. The surface area of the concert hall will be $30^2 = 900$ times as great as the surface area of the model (because areas always scale with the square of the scale factor).

65. The height of the office complex will be 80 times as large as the height of the model.

67. You would need $80^3 = 512{,}000$ times as many marbles to fill the office complex (volumes always scale with the cube of the scale factor).

69. Your waist size is a measurement of length, so it has increased by a factor of 4.

71. Your weight is a function of your volume, so it has increased by a factor of $4^3 = 64$.

73. Answers will vary. Example: If your waist size is 32 inches, Sam's is $1.2 \times 32 \text{ in} = 38.4$ inches.

75. Think of a soup can (a squirrel) and a trash can (a human), compute their surface areas and volumes, find their surface-area-to-volume ratios, and you will discover that the surface-area-to-volume ratio for squirrels is larger than that of humans. Another way to understand this fact: surface area scales with the square of the scale factor, whereas volume scales with the cube of the scale factor. Thus the surface-area-to-volume ratio decreases when objects are scaled up (and in fact, it scales with the reciprocal of the scale factor).

77. The moon's surface-area-to-volume ratio is four (i.e. the scale factor) times as large as Earth's, (see discussion at the end of Exercise 75).

79. If you understood the comment at the end of Exercise 75, the surface-area-to-volume ratio of the bowling ball should be 1/3 the ratio of the softball, because the bowling ball is 3 times as large. Computing directly: The surface area of the softball is $4\pi \times (2 \text{ in})^2 = 50.3 \text{ in}^2$, its volume is $\frac{4}{3}\pi \times (2 \text{ in})^3 = 33.5 \text{ in}^3$, and its surface-area-to-volume ratio is 1.5. The surface area of the bowling ball is $4\pi \times (6 \text{ in})^2 = 452.4 \text{ in}^2$, its volume is $\frac{4}{3}\pi \times (6 \text{ in})^3 = 904.8 \text{ in}^3$, and its surface-area-to-volume ratio is 0.5. As expected, the ratio of the bowling ball is 1/3 the ratio of the softball.

FURTHER APPLICATIONS

81. a. The book occupies three-dimensional space, so three dimensions are needed to describe it.
    b. The cover of the book occupies a portion of a plane, so two dimensions are needed to describe it.
    c. An edge of the book lies along a line, so a single dimension is adequate for its description.
    d. A point is dimensionless, so anything on the book that corresponds to a point (such as a corner of a page) is of zero dimension.

83. The third line is necessarily perpendicular to both lines due to a theorem from geometry (when two parallel lines are cut by any other line, all three lying in the same plane, alternate interior angles are equal – this amounts to saying that if the angle between one of the parallel lines and the third line is 90°, the angle between the other parallel line and the third line is 90°).

85. a. The surface area of an individual sac is $4\pi \times \left(\frac{1}{6} \text{ mm}\right)^2 = 0.349 \text{ mm}^2$, and the total surface area is 300 million times this value, which is $1.05 \times 10^8 \text{ mm}^2 = 105 \text{ m}^2$. The volume of a single sac is $\frac{4}{3}\pi \times \left(\frac{1}{6} \text{ mm}\right)^3 = 0.0194 \text{ mm}^3$, so the total volume is $5.82 \times 10^6 \text{ mm}^3$ (multiply the individual volume by 300 million).

    b. We need to solve $5.82 \times 10^6 = \frac{4}{3}\pi \times r^3$ for r to find the radius of such a sphere. Multiply both sides of the equation by $\frac{3}{4\pi}$ and take the cube root to find $r = 112$ mm (rounded). This sphere has surface area of $4\pi \times (111.572 \text{ mm})^2 = 1.56 \times 10^5 \text{ mm}^2$. The surface area of the air sacs is $1.05 \times 10^8 \text{ mm} \div 1.56 \times 10^5 \text{ mm} = 673$ times as large as the surface area of the hypothetical sphere.

    c. We need to solve $1.05 \times 10^8 = 4\pi \times r^2$ for r to find the radius of such a sphere. Divide both sides by 4π and take the square root to find $r = 2891$ mm, or about 2.9 meters (i.e. more than 9 feet). The human lung has a remarkable design as it is able to take advantage of large surface area despite its relatively small volume.

87. The volume of one tunnel is $\frac{1}{2}\pi\times(4\text{ m})^2\times 50{,}000\text{ m}=1.26\times10^6\text{ m}^3$, so the volume of all three is $3.8\times10^6\text{ m}^3=0.0038\text{ km}^3$. Depending on whether the dimensions given are for a rough tunnel, or the structure that a motorist sees driving through it, the amount of earth removed may be considerably more than the answer provided.

UNIT 10B: PROBLEM SOLVING WITH GEOMETRY

QUICK QUIZ

1. **c. Each degree contains 60 minutes.**
2. **b. Each degree contains 60 minutes, and each minute contains 60 seconds.**
3. **a. Lines of latitude run east and west on the globe, parallel to the equator.**
4. **b. That you are at 30°S means you are in the southern hemisphere (and thus not in North America). A check of the globe will reveal these coordinates lie in the south Pacific.**
5. **b. Angular size is a function of the distance from the object: when the distance from an object increases, its angular size decreases.**
6. **b. Imagine a solar eclipse, when the Moon is positioned between the Earth and the Sun, and think of the triangle that stretches from a point on Earth to the edges of the moon. Imagine another triangle that continues beyond the Moon to the edges of the Sun (a simple drawing will help you to visualize the situation). These two triangles are similar, and thus we can write the ratio $\frac{\text{diameter of Sun}}{\text{diameter of Moon}}=\frac{400x}{x}$, where x represents the distance from the Earth to the Moon. This ratio implies that the diameter of the Sun is 400 times the diameter of the Moon. One can also appeal to the angular size formula in the text to get the same result.**
7. **a. A 10% grade means the slope is 1/10, which implies a vertical change of 1 unit for every horizontal change of 10 units.**
8. **a. The Pythagorean theorem gives $x=\sqrt{6^2+9^2}$.**
9. **a. The statement in answer a is an application of the fact that the ratio of corresponding sides in a pair of similar triangles is equal to the ratio of a different set of corresponding sides.**
10. **b. As explained in Example 9 of the text, a circle is the solution to the problem of enclosing the largest possible area with a fixed perimeter.**

REVIEW QUESTIONS

1. **Each degree is divided into 60 minutes and each minute is divided into 60 seconds The symbols ′ and ″ are used for minutes and seconds, respectively.**

3. **Angular size is the angle that an object covers as seen from your eye. The farther away an object is located from you, the smaller it will appear to you. For angular sizes less than a few degrees:**

$$\text{angular size}=\text{physical size}\times\frac{360^\circ}{2\pi\times\text{distance}}.$$

5. **The Pythagorean theorem can be used to measure the heights of buildings or the length of diagonal trusses in buildings.**

7. **Similar triangles can be used to determine the heights of buildings by comparing the shadow lengths of an object of known height with the shadow length of the building in question.**

DOES IT MAKE SENSE?

9. **Does not make sense. Points south of the equator experience summer in December.**
11. **Makes sense. As long as the speaker can ride a bike, a 7% grade should pose no difficulty.**
13. **Makes sense. Triangle B is just a scaled up version of Triangle A, so they are similar.**

BASIC SKILLS AND CONCEPTS

15. $32.5° = 32° + 0.5° \times \frac{60'}{1°} = 32°30'0''$

17. $12.33° = 12° + 0.33° \times \frac{60'}{1°} = 12°19.8' = 12° + 19' + 0.8' \times \frac{60''}{1'} = 12°19'48''$

19. $149.83° = 149° + 0.83° \times \frac{60'}{1°} = 149°49.8' = 149° + 49' + 0.8' \times \frac{60''}{1'} = 149°49'48''$

21. $30°10' = 30° + 10' \times \frac{1°}{60'} = 30.17°$

23. $123°10'36'' = 123° + 10' \times \frac{1°}{60'} + 36'' \times \frac{1°}{3600''} = 123.18°$

25. $8°59'10'' = 8° + 59' \times \frac{1°}{60'} + 10'' \times \frac{1°}{3600''} = 8.99°$

27. There are $360° \times \frac{60'}{1°} = 21{,}600'$ in a circle.

29. Madrid is at 40°N, 4°W.

31. The latitude changes from 44°N to 44°S. To find the longitude, move 180° east from 79°W to arrive at 101°E. (Note that after moving 79°, you'll be at the prime meridian, or 0°, and you still have 101° to go). Thus the point opposite Toronto is 44°S, 101°E.

33. Buenos Aires is farther from the North Pole because its latitude (35°S) is further south than the latitude of Capetown (34°S).

35. Buffalo is 17° north of Miami, and each degree is 1/360 of the circumference of the Earth. Since the circumference is about 25,000 mi, Buffalo is $\frac{17}{360} \times 25{,}000 = 1181$ mi, or about 1200 miles from Miami.

37. A quarter is about an inch in diameter, and 3 yards is 108 inches. Thus its angular size is $1 \text{ in} \times \frac{360°}{2\pi \times 108 \text{ in}} = 0.53°$.

39. The Sun's true diameter (physical size) is $0.5° \times \frac{2\pi \times 150{,}000{,}000 \text{ km}}{360°} = 1.31 \times 10^6$ km.

41. A roof with a pitch of 1 in 4 has a slope of 1/4, which is steeper than a roof with a slope of 2/10.

43. A railroad with a 3% grade has a slope of $3/100 = 0.03$, which is not as steep as a railroad with a slope of $1/25 = 0.04$.

45. The slope of a 8 in 12 roof is 8/12. The roof will rise $\frac{8}{12} \times 15 \text{ ft} = 10$ feet in a horizontal run of 15 feet.

47. Think of a right triangle with side lengths of 6 and 6 – this produces an isosceles triangle, which implies the non-right angles are both 45°. It is possible to have a 7 in 6 roof: it's just a steep roof that rises 7 feet for every 6 feet of horizontal run.

49. The slope of such a road is $20/150 = 0.133$, so its grade is 13.3%.

51. a. Walk 6 blocks east (6/8 mi) and 1 block north (1/5 mi) for a total distance of $19/20 = 0.95$ mi.

b. The Pythagorean theorem gives the straight-line distance as $\sqrt{(6/8)^2 + (1/5)^2} = 0.78$ mi.

53. a. Walk 2 blocks west (2/8 mi) and 3 blocks north (3/5 mi) for a total distance of $17/20 = 0.85$ mi.

 b. The Pythagorean theorem gives the straight-line distance as $\sqrt{(2/8)^2 + (3/5)^2} = 0.65$ mi.

55. a. Walk 3 blocks east (3/8 mi) and 3 blocks south (3/5 mi) for a total distance of $39/40 = 0.98$ mi.

 b. The Pythagorean theorem gives the straight-line distance as $\sqrt{(3/8)^2 + (3/5)^2} = 0.71$ mi.

57. The height of the triangle in Figure 10.33 is $\sqrt{(800 \text{ ft})^2 - (200 \text{ ft})^2} = 774.6$ ft, and thus the area of the lot is $\frac{1}{2} \times 200 \text{ ft} \times 774.6 \text{ ft} = 77{,}460 \text{ ft}^2$, which is $77{,}460 \text{ ft}^2 \times \frac{1 \text{ acre}}{43{,}560 \text{ ft}^2} = 1.78$ acres.

59. The height of the triangle in Figure 10.33 is $\sqrt{(3800 \text{ ft})^2 - (600 \text{ ft})^2} = 3752.3$ ft, and thus the area of the lot is $\frac{1}{2} \times 600 \text{ ft} \times 3752.3 \text{ ft} = 1{,}125{,}700 \text{ ft}^2$, which is $1{,}125{,}700 \text{ ft}^2 \times \frac{1 \text{ acre}}{43{,}560 \text{ ft}^2} = 25.84$ acres.

61. The larger triangle is just a scaled up version of the smaller triangle (or so it appears), and so they are similar.

63. The triangle on the left appears to be an isosceles triangle, whereas the one on the right does not, which means the triangles don't have the same angle measure, and cannot be similar.

65. $8/10 = x/5$, which implies $x = 4$. Use the Pythagorean theorem to show that $y = 3$ (we know the triangles are right because the left triangle satisfies $6^2 + 8^2 = 10^2$).

67. $x/60 = 10/40$, which implies $x = 15$. $50/60 = y/40$, which implies $y = 100/3 = 33.3$.

69. Refer to Figure 10.36, and note that a 12-ft fence that casts a 25-foot shadow is equivalent to a house of height h (set back 60 feet from the property line) that casts an 85-ft shadow. We can write the ratio $12/25 = h/85$, which implies $h = 40.8$ ft.

71. As in Exercise 69, we can write $12/30 = h/80$, which implies $h = 32$ ft.

73. The radius of a circle with circumference of 50 m is $\frac{50 \text{ m}}{2\pi} = 7.96$ m, so its area is $\pi \times (7.96 \text{ m})^2 = 199 \text{ m}^2$.

 The side length of a square with perimeter of $50 \text{ m} \div 4 = 12.5$ m, so its area is $12.5 \text{ m} \times 12.5 \text{ m} = 156 \text{ m}^2$. The area of the circular region is larger.

75. The radius of a circle with circumference of 150 m is $\frac{150 \text{ m}}{2\pi} = 23.9$ m, so its area is $\pi \times (23.9 \text{ m})^2 = 1795 \text{ m}^2$.

 The side length of a square with perimeter of 150 m is $150 \text{ m} \div 4 = 37.5$ m, so its area is $37.5 \text{ m} \times 37.5 \text{ m} = 1406.25 \text{ m}^2$. The area of the circular region is larger.

77. For the first can described, the area of the top and bottom is $2 \times \pi \times (4 \text{ in})^2 = 100.53 \text{ in}^2$, and the area of the side is $2\pi \times 4 \text{ in} \times 5 \text{ in} = 125.66 \text{ in}^2$. The cost of the can is $100.53 \text{ in}^2 \times \frac{\$1.00}{\text{in}^2} + 125.66 \text{ in}^2 \times \frac{\$0.50}{\text{in}^2} = \$163.36$.

 Using similar calculations, the cost of the second can is $157.08 \text{ in}^2 \times \frac{\$1.00}{\text{in}^2} + 125.66 \text{ in}^2 \times \frac{\$0.50}{\text{in}^2} = \$219.91$.

79. Assuming a rectangular box, the most economical shape is a cube, with side length equal to 2 ft (see Example 10 in the text). Such a box has 6 faces, each with area of 4 ft^2, for a total area of 24 ft^2. At \$0.15 per square foot, the box would cost \$3.60.

FURTHER APPLICATIONS

81. a. The area of the storage region of the Blu-ray disc, which is the area of the entire Blu-ray disc minus the area of the inner circle (that is not part of the storage region), is $\pi \times (5.9\text{ cm})^2 - \pi \times (2.5\text{ cm})^2 = 89.7\text{ cm}^2$.

b. The density is $\dfrac{50{,}000\text{ million bytes}}{89.7\text{ cm}^2} = 557\text{ million bytes/cm}^2$.

c. The length of the groove is $\dfrac{\pi\left((5.9\text{ cm})^2 - (2.5\text{ cm})^2\right)}{0.3\text{ micrometer}} \times \dfrac{10^6\text{ micrometer}}{1\text{ m}} \times \dfrac{1\text{ m}}{100\text{ cm}} = 2{,}990{,}796\text{ cm}$, which is $2{,}990{,}796\text{ cm} \times \dfrac{1\text{ m}}{100\text{ cm}} \times \dfrac{1\text{ km}}{1000\text{ m}} \times \dfrac{1\text{ mi}}{1.6093\text{ km}} = 18.6\text{ mi}$.

83. The throw goes along the hypotenuse of a right triangle, whose length is $\sqrt{(90\text{ ft})^2 + (90\text{ ft})^2} = 127.3\text{ ft}$.

85. For simplicity, assume you can row at a rate of 1.0 mph, and you can bike at a rate of 1.5 mph. (The following calculations would yield the same result for any chosen rowing rate.) Using $\text{time} = \dfrac{\text{distance}}{\text{rate}}$, the time it takes to bike along the edges of the reservoir is $\dfrac{1.2\text{ mi} + 0.9\text{ mi}}{1.5\text{ mph}} = 1.4\text{ hr}$. The time it takes to row along the hypotenuse is $\dfrac{\sqrt{(1.2\text{ mi})^2 + (0.9\text{ mi})^2}}{1.0\text{ mph}} = 1.5\text{ hr}$, so biking is faster.

87. a. The volume of the water in the bed is $8\text{ ft} \times 7\text{ ft} \times 0.75\text{ ft} = 42\text{ ft}^3$. Divide this volume by the area of the lower room (80 ft^2) to find that the depth of the water is 0.525 ft.

b. The weight of the water in the bed is $42\text{ ft}^3 \times \dfrac{62.4\text{ lb}}{1\text{ ft}^3} = 2621\text{ lb}$.

89. The perimeter of a corral with dimensions of 10 m by 40 m is 100 m. One could use a square-shaped corral with side length of 20 m to achieve the desired area of 400 square meters; this corral would have a perimeter of 80 m, and would require less fencing.

91. Your project will cost $3\text{ mi} \times \dfrac{\$500}{1\text{ mi}} + \sqrt{(2\text{ mi})^2 + (1\text{ mi})^2} \times \dfrac{\$1000}{1\text{ mi}} = \$3736$. The plan suggested by your boss will cost $4\text{ mi} \times \dfrac{\$500}{1\text{ mi}} + \sqrt{(1\text{ mi})^2 + (1\text{ mi})^2} \times \dfrac{\$1000}{1\text{ mi}} = \$3414$.

93. a. With a lot of work and patience, you might be able to come up with the dimensions of the can that has the lowest surface area (and thus lowest cost): $r = 3.8$ cm and $h = 7.7$ cm. One way to save time is to observe that $355 = \pi r^2 h$, which implies that $h = \dfrac{355}{\pi r^2}$. Insert this expression for h into the surface area formula $2\pi rh + 2\pi r^2$ to find that the surface area can be expressed as a function of r: $A = \dfrac{710}{r} + 2\pi r^2$. Since the cost is directly proportional to the surface area, the object is to find the value of r that yields the smallest surface area A. If you graph A on a calculator, and use the trace function, you'll see that $r = 3.8$ at the low point of the graph.

b. The dimensions of the optimal can found above imply the can is as wide (diameter) as it is high. Soda cans aren't built that way, probably for a variety of reasons. (Among them: the can needs to fit into the hand of an average human, and the top and bottom of the can cost more because the aluminum is thicker).

95. a. Note that the radius of the cone will be 6 ft (because the height is 1/3 the radius). The volume is $\frac{1}{3}\pi \times (6 \text{ ft})^2 \times 2 \text{ ft} = 75.4 \text{ ft}^3$.

b. Again, note that $r = 3h$, and solve $1000 \text{ ft}^3 = \frac{1}{3}\pi \times (3h)^2 \times h = 3\pi \times h^3$ for h to find $h = 4.7$ ft.

c. Assuming 10,000 grains per cubic inch, there are $75.4 \text{ ft}^3 \times \left(\frac{12 \text{ in}}{1 \text{ ft}}\right)^3 \times \frac{10{,}000 \text{ grains}}{\text{in}^3} = 1.3$ billion grains. The figure of 10,000 grains per cubic inch is very much dependent upon the size of sand grain assumed, which varies.

97. a. The pyramid's height is $481 \text{ yd} \div 3 = 160.3$ yards, which is about 1.6 times the length of a football field (excluding the end zones).

b. The volume of the pyramid is $\frac{1}{3} \times (756 \text{ ft})^2 \times 481 \text{ ft} = 91{,}636{,}272 \text{ ft}^3$, which is $91{,}636{,}272 \text{ ft}^3 \times \left(\frac{1 \text{ yd}}{3 \text{ ft}}\right)^3 = 3{,}393{,}936 \text{ yd}^3$.

c. Divide the result in part (b) by 1.5 to get around 2,263,000 blocks.

d. $2{,}263{,}000 \text{ blocks} \times \frac{2.5 \text{ min}}{\text{block}} \times \frac{1 \text{ hr}}{60 \text{ min}} \times \frac{1 \text{ d}}{12 \text{ hr}} \times \frac{1 \text{ yr}}{365 \text{ d}} = 21.5$ years. This is comparable to the amount of time suggested in historical records, which indicates that Lehner's estimate of 10,000 workers is in error. However, we don't have all the facts – perhaps Lehner began with the assumption that, with a work force of 10,000 laborers, one stone could be placed every 2.5 minutes, or maybe he assumed laborers worked on the project 24 hours per day.

e. The volume of the Eiffel tower is $\frac{1}{3} \times (120 \text{ ft})^2 \times 980 \text{ ft} = 4{,}704{,}000 \text{ ft}^3$, which is about 5% of the pyramid's volume.

UNIT 10C: FRACTAL GEOMETRY

QUICK QUIZ

1. b. As noted in the text, fractals successfully replicate natural forms, especially when random iteration is used.
2. a. A coastline can be modeled with fractal curves which have dimension between 1 and 2, and by the very definition of fractal dimension, we see that shortening the length of the ruler increases the number of elements (which, in turn, represents the length of the coastline).
3. b. The edge of a leaf (think of a maple leaf) has many of the properties of a fractal, such as self similarity, and a length that increases as one decreases the size of the ruler used to measure it.
4. c. Fractals often have the property of self similarity, which means that under greater magnification, the underlying pattern repeats itself.
5. c. The definition of the fractal dimension of an object leads to fractional values for the dimension.
6. c. Data suggest that most coastlines have a fractal dimension of about 1.25.
7. a. The curve labeled L_6 in Figure 10.55 is simply the sixth step (or iteration) of the process required to generate the snowflake curve. The snowflake curve itself is denoted by L_∞ (the end product of performing infinite iterations).
8. c. An area bounded by a finite curve (it is understood that by this we mean a closed curve of finite length) cannot possibly be of infinite area. If it were possible, we'd have to throw all of our notions about the areas of bounded plane regions out the door. How could the area of a region contained within a region of a finite area be infinitely large?

9. **a.** The definition of a self-similar fractal is that its patterns repeat themselves under greater magnification.

10. **b.** From previous units, we know that if you take, say, an ice cube, and break it into smaller pieces, the aggregate surface area of the pieces is larger than the surface area of the original cube of ice. The same idea is illustrated by the process of creating the Sierpinski sponge. In the first step, when the four subcubes are removed, it can be shown that the surface area increases by a factor of 4/3. In fact, this happens in every step, and thus the surface area becomes infinitely large as the process continues.

REVIEW QUESTIONS

1. Objects, like a coastline, that continually reveal new features at smaller scales are called fractals. Using a smaller ruler to measure the perimeter of a fractal allows more of the actual edge to be measured, leading to a larger calculated perimeter.

3. R represents the reduction factor, the factor by which the measuring unit is reduced. N represents the factor by which the number of measuring units increases when the measuring unit is reduced by R.

5. The snowflake island is a region bounded by three snowflake curves. Snowflake islands are created by starting with an equilateral triangle. Each of the three sides of the triangle are then turned into a snowflake curve. (See Figure 10.56 on page 613.) Since single snowflake curves are infinitely long, the coastline of the snowflake island, consisting of three snowflake curves, will also be infinitely long. Since the size of the snowflake island does not grow without bound, the snowflake island is an object with a finite area and an infinitely long boundary.

7. Iteration is the process of repeatedly applying a rule over and over to generate a self-similar fractal. The Cantor set is produced by starting with a line segment and iterating with the following rule: Delete the middle third of each line segment of the current figure. The fractal dimension of the Cantor set is less than 1.
   The Sierpinski triangle is produced by starting with a solid black equilateral triangle and iterating with the following rule: For each black triangle in the current figure, connect the midpoints of the sides and remove the resulting inner triangle. The fractal dimension of the Sierpinski triangle is between 1 and 2.
   The Sierpinski sponge is produced by starting with a solid cube and iterating with the following rule: Divide each cube of the current object into 27 identical subcubes and remove the central subcube and the center cube of each face. The resulting object has a fractal dimension between 2 and 3.

DOES IT MAKE SENSE?

9. Makes sense. The boundary of a rectangle is made of straight lines whose lengths are well defined, and so can be measured with a standard ruler (to a reasonable degree of accuracy).

11. Does not make sense. The boundary of the snowflake island is infinitely long, and though its area is finite, it is not computed with a simple "length times width" formula.

13. Makes sense. Data have been collected that shows many boundaries in nature, such as coastlines and edges of leaves, have fractal dimensions greater than one.

BASIC SKILLS AND CONCEPTS

15. According to the definition of fractal dimension, $R=2$ and $N=2$, so we have $2=2^D$, where D is the fractal dimension. It is evident that $D=1$ in this case, and thus the object is not a fractal (it behaves like an ordinary geometric object with regard to length measurements).

17. According to the definition of fractal dimension, $R=2$ and $N=8$, so we have $8=2^D$, where D is the fractal dimension. It is evident that $D=3$ in this case, and thus the object is not a fractal (it behaves like an ordinary geometric object with regard to volume measurements).

19. According to the definition of fractal dimension, $R=2$ and $N=6$, so we have $6=2^D$, where D is the fractal dimension. Solving for D using logarithms produces a value of $D=\frac{\log_{10} 6}{\log_{10} 2}=2.585$, and thus the object is a fractal (it does not behave like an ordinary geometric object with regard to area measurements).

21. According to the definition of fractal dimension, $R = 5$ and $N = 5$, so we have $5 = 5^D$, where D is the fractal dimension. It is evident that $D = 1$ in this case, and thus the object is not a fractal (it behaves like an ordinary geometric object with regard to length measurements).

23. According to the definition of fractal dimension, $R = 5$ and $N = 125$, so we have $125 = 5^D$, where D is the fractal dimension. It is evident that $D = 3$ in this case, and thus the object is not a fractal (it behaves like an ordinary geometric object with regard to volume measurements).

25. According to the definition of fractal dimension, $R = 5$ and $N = 30$, so we have $30 = 5^D$, where D is the fractal dimension. Solving for D using logarithms produces a value of $D = \frac{\log_{10} 30}{\log_{10} 5} = 2.113$, and thus the object is a fractal (it does not behave like an ordinary geometric object with regard to area measurements).

27. a. Suppose that the original line segment has length of 1 unit. When we measure the length of the first iteration of the quadratic Koch curve with a ruler of length 1 unit (or one element), we find the length is 1 unit, the distance from one endpoint to the other. When we shorten the ruler by a factor of $R = 4$ so that its length is 1/4 unit, the length of the first iteration is $N = 8$ elements (i.e., the length is $8/4 = 2$ units). This is true at every stage of the process.

b. According to the definition of fractal dimension, we have $8 = 4^D$. Solving for D yields $D = \frac{\log_{10} 8}{\log_{10} 4} = 1.5$.

The length of the quadratic Koch curve is infinite, because at each iteration, the length increases by a factor of 2, and thus its length grows without bound as the process continues.

c. Consider what happens to the area of the quadratic Koch island as we go from the first stage (a square) to the second stage. Along the upper boundary, the area is increased by a small square drawn above the boundary, but it is decreased by a small square drawn below the boundary. This is true along every edge of the original square, and the net effect is no change in the area. The same is true for every step of the process, so the area of the final island is equal to the area of the original square. The coastline of the island is infinite, as noted above in part (b).

FURTHER APPLICATIONS

29. When the ruler is reduced by a factor of $R = 9$, there will be $N = 4$ elements found (each of length 1/9). This gives $4 = 9^D$, which when solved leads to $D = \frac{\log_{10} 4}{\log_{10} 9} = 0.631$. (Note that this is the answer you get no matter what stage of the process you choose to analyze). The dimension is less than 1 due to the fact that the end result of the process of constructing the Cantor set is a set of isolated points. Though there are infinitely many points, there aren't "enough" of them to constitute a line segment of measurable length (that is, this fractal object does not behave like an ordinary geometric object with regard to length).

31. a. Note that every time the ruler is decreased in length by a factor of $R = 10$, the number of elements increases by a factor of $N = 20$, which leads to $20 = 10^D$. This implies the fractal dimension is $D = \frac{\log_{10} 20}{\log_{10} 10} = 1.301$.

b. Note that every time the ruler is decreased in length by a factor of $R = 2$, the number of area elements increases by a factor of $N = 3$, which leads to $3 = 2^D$. This implies the fractal dimension is $D = \frac{\log_{10} 3}{\log_{10} 2} = 1.585$. The fractal dimension is less than two because the standard notion of surface area for ordinary objects does not carry over to the surface area of fractal objects.

31. (continued)

c. Note that every time the ruler is decreased in length by a factor of $R = 2$, the number of volume elements increases by a factor of $N = 6$, which leads to $6 = 2^D$. This implies the fractal dimension is $D = \frac{\log_{10} 6}{\log_{10} 2} = 2.585$. A fractal dimension between 2 and 3 is reasonable because such a rock exhibits the same properties as the Sierpinski sponge – it is somewhat "less" than a solid three-dimensional cube because material has been removed in a fractal pattern.

33. The branching in many natural objects has the same pattern repeated on many different scales. This is the process by which self-similar fractals are generated. Euclidean geometry is not equipped to describe the repetitions of patterns on many scales.

UNIT 11A: MATHEMATICS AND MUSIC

QUICK QUIZ

1. **b.** Instruments with strings (guitar, violin, piano), woodwinds with reeds (clarinet, bassoon), and instruments with a column of air in a tube (organ pipe, horn, flute) – all produce their sound with an object that vibrates.

2. **a.** In the case of a string vibrating, a single cycle of vibration corresponds to the string moving to its high point and to its low point, so 100 cps means the string is at its high point (and low point) 100 times each second.

3. **a.** Higher pitches go hand in hand with higher frequencies.

4. **c.** The fundamental frequency of a string occurs when the string vibrates up and down along its entire length, and this produces the lowest possible pitch from that string. (Though the text does not discuss it, you may know that a longer wavelength produces a lower frequency – the wavelength can't get any longer for a particular string than the wave associated with the fundamental frequency).

5. **a.** Every time you raise the pitch of a sound by an octave, the pitch doubles.

6. **c.** The 12-tone scale uses twelve equally spaced (in the sense described below) notes for every octave. The factor f by which the frequency changes in moving from one note to the next is constant, which means that in multiplying an initial frequency by f 12 times in a row, you've moved an octave up the scale, and doubled the frequency. This gives us the relationship $f^{12} = 2$, which in turn implies that $f = \sqrt[12]{2}$.

7. **c.** Table 11.1 shows the frequencies of each of the notes in a particular octave; the names of the notes repeat every octave. As you can see in the table, increasing the frequency of middle C by a factor of 1.5 moves you to a G, so if you increase the frequency of the next higher C by a factor of 1.5, you'll end up at the next higher G.

8. **b.** The frequency of each note in the 12-tone scale is related to the note one half-step below it by a multiplicative factor of $f = \sqrt[12]{2} = 1.05946$. This means the frequencies increase by about 5.9% every half-step, which is exponential growth.

9. **b.** The mathematician Fourier discovered that musical sounds are the sum of several constant-frequency waves.

10. **a.** The process of digitizing analog sound changes wave forms that represent sound waves into lists of numbers that can be stored on CDs, and reinterpreted by a CD player to produce a musical sound.

REVIEW QUESTIONS

1. Pitch is a basic quality of sound. Bass notes have lower pitches while treble and alto notes have increasingly higher pitches. The higher the frequency, the higher the pitch.

3. In a 12-tone scale, there are twelve notes in each octave, called half-steps. Moving from one note to the next in the scale increases the frequency of the note by a factor of $\sqrt[12]{2} \approx 1.05946$.

5. Real wave forms consist of a combination of simple waves that are the harmonics of the fundamental. They are far richer in tone than simple wave forms.

DOES IT MAKE SENSE?

7. Does not make sense. The pitch of a string is a function of the length of the string, not the number of times the string is plucked.

9. Makes sense. The frequency of each note in the 12-tone scale increases by a factor of $f = \sqrt[12]{2} = 1.05946$, and this is exponential growth.

BASIC SKILLS AND CONCEPTS

11. The next higher octave corresponds to a doubling in frequency, so the requested frequencies are 440 cps, 880 cps, 1760 cps, and 3520 cps.

13.

Note	Frequency (cps)
F	347
F#	368
G	389
G#	413
A	437
A#	463
B	491
C	520
C#	551
D	584
D#	618
E	655

The table can be generated beginning with the initial frequency of 347 cps, and multiplying by $f=\sqrt[12]{2}$ to produce each successive note. Realize that you will get slightly different values for some of the entries in the table if you generate it from middle C, whose frequency is 260 cps. This is due to rounding errors, and it explains the different values shown in this table and in Table 11.1 at the middle G entry. We began with an initial value of 347 cps to generate this table, but that's a rounded value for the true frequency of middle F.

15. a. $260\times\left(\sqrt[12]{2}\right)^7=390$ cps

b. $260\times\left(\sqrt[12]{2}\right)^9=437$ cps

c. An octave is 12 half-steps, so use a total of 19 half-steps. $260\times\left(\sqrt[12]{2}\right)^{19}=779$ cps.

d. $260\times\left(\sqrt[12]{2}\right)^{25}=1102$ cps

e. An octave is 12 half-steps, so use a total of 39 half-steps. $260\times\left(\sqrt[12]{2}\right)^{39}=2474$ cps.

17. To compute the frequency for a note one half-step below a particular frequency, we simply divide by $f=\sqrt[12]{2}$, which is equivalent to multiplying by the factor f raised to the power of –1. Thus to find the note 7 half steps below middle A, use $437\times\left(\sqrt[12]{2}\right)^{-7}=292$ cps. Ten half-steps below middle A is $437\times\left(\sqrt[12]{2}\right)^{-10}=245$ cps.

FURTHER APPLICATIONS

19. a. $\left(\sqrt[12]{2}\right)^7=2^{7/12}$.

b. From part (a), it raises by a factor of $\left(2^{7/12}\right)^2=2^{14/12}=2.245$.

c.

Note	Frequency (cps)	Note	Frequency (cps)
C	260	C	260
G	390	G	390
D	584	D	584
A	875	A	875
E	1310	E	1310
B	1963	B	1963
F#	2942	F#	2942
C#	4407	C#	4407
G#	6604	G#	6604
D#	9894	D#	9894
A#	14,825	A#	14,825
F	22,212	F	22,212
C	33,280	C	33,280

19. (continued)

The table can be generated beginning with the initial frequency of 260 cps, and multiplying by a factor of $2^{7/12}$ to produce successive notes.

d. As shown in the table in part (c), you need 12 fifths to return to a C, which is 7 octaves. You can also understand this by analyzing the factor $2^{7/12}$.

e. The ratio is 33,280/260 = 118. This, too, can be understood in the context of the multiplicative factor $2^{7/12}$ used to raise a note by a fifth. If you use this factor 12 times, you will have increased the frequency of the initial note by a factor of $\left(2^{7/12}\right)^{12} = 2^7 = 128$. Another way to look at it: move one octave up, and you double the frequency. Here we have moved up 7 octaves, so the frequency has increased by a factor of 2^7.

21. There are 2 half notes in a 4/4-time measure, 8 eighth notes, and 16 sixteenth notes.

UNIT 11B: PERSPECTIVE AND SYMMETRY

QUICK QUIZ

1. **b.** The principal vanishing point in a painting is the point of intersection of those lines in the painting that are parallel in the real scene, and perpendicular to the canvas. We are assuming here that the tracks are, in fact, perpendicular to the canvas, and that they are straight.
2. **b.** The *horizon line* is the line (horizontal, of course) through the principal vanishing point. According to the principals of perspective, all sets of parallel lines in the real scene intersect at some point along the horizon line in the painting.
3. **a.** Not only does the painting show this fact, but the parallel beams in the ceiling are perpendicular to the canvas, so they should intersect at the principal vanishing point.
4. **c.** There is a vertical line of symmetry in the center of da Vinci's sketch, as both sides of the drawing are nearly identical on either side of this line.
5. **c.** If the letter **W** were reflected across this line, it would appear unchanged from the original.
6. **b.** Rotate the letter **Z** through 180°, and it remains unchanged in appearance.
7. **a.** Because a circle can be rotated through any angle and remain unchanged, it has rotation symmetry. It also has reflection symmetry over any line through its center, for the same reason.
8. **c.** The only regular polygons that admit complete tilings are equilateral triangles, squares, and regular hexagons.
9. **a.** Refer to Figure 11.23 in the text.
10. **c.** A periodic tiling is one where a pattern repeats itself throughout the tiling.

REVIEW QUESTIONS

1. Perspective is used to create the appearance of three dimensions in a two dimensional image by mimicking how parallel lines appear to the human eye as they recede into the distance. In mathematics, symmetry is a property of an object that remains unchanged under certain operations, most commonly, reflection, translation, and rotation.

3. The horizon line is the horizontal line passing through the principal vanishing point. All sets of lines that are mutually parallel in the real scene (except those parallel to the horizon line) must meet at their own vanishing point on the horizon line in order to maintain the correct perspective of the entire scene.

5. Tilings are arrangements of polygons, covering a flat area, that interlock perfectly without overlapping.

7. If the restriction on using regular polygons is removed, more tilings are possible since other polygons can be used where their interior angles evenly divide 360°.

DOES IT MAKE SENSE?

9. Does not make sense. The principal vanishing point is often very apparent in paintings that are rendered in proper perspective (see Figures 11.6 and 11.8, for example).

11. Makes sense. It is the science and art of perspective painting that allows talented artists to render realistic three-dimensional scenes on a two-dimensional canvas.

13. Does not make sense. Only equilateral triangles, squares, and regular hexagons allow for complete tilings. Susan should rent a tile saw.

BASIC SKILLS AND CONCEPTS

15. a. The only obvious vanishing point is depicted in the diagram below. It is not the principal vanishing point because the lines of the road do not meet the canvas at right angles. However, it does lie on the horizon line.

    b.

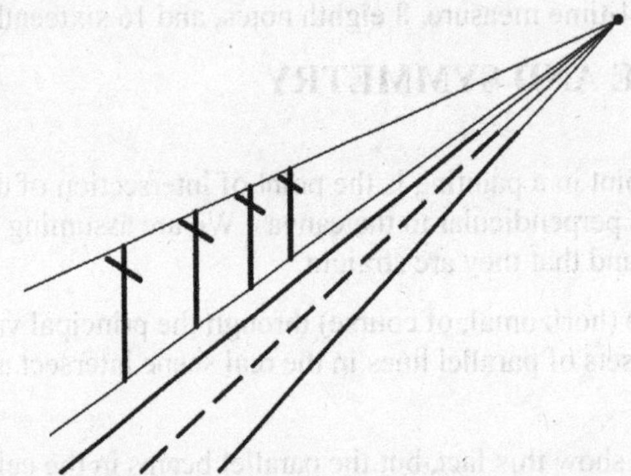

17.

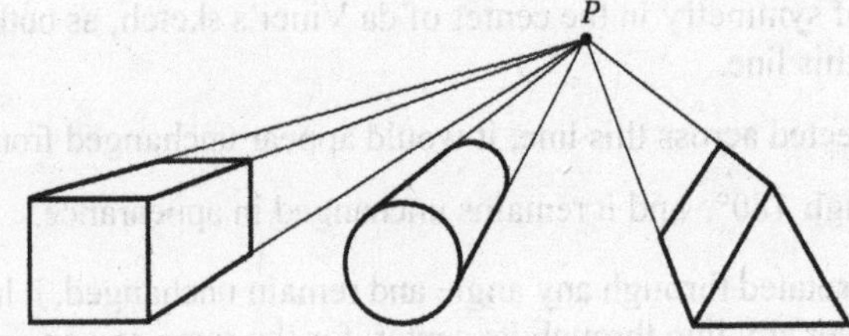

19. a. In order to draw the equally spaced poles, first draw a perspective line along the bases of the existing poles, and draw one along the tops. These two lines should intersect at a vanishing point (directly on, or very near to the horizon). Now measure 2.0 cm along the base line, and draw a vertical pole to meet the top perspective line. Repeat this procedure for the second pole.

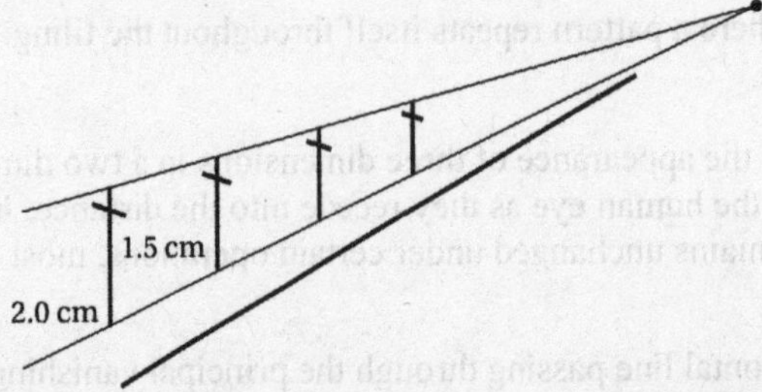

    b. The heights of the poles can be measured directly with a ruler. You should find the first pole you drew is about 1.0 cm, and the second about 0.5 cm. Notice that the lengths of the poles are decreasing in a linear fashion (2.0, 1.5, 1.0, and 0.5 cm). This is due to the fact that the poles are sandwiched between two lines of constant slope, and that the distances between them are equal. (Also note that the diagram in part (a) is not to scale).

19. (continued)

  c. If your drawing is an accurate depiction of the scene, the poles are not equally spaced in the actual scene. This is due to perspective: equal distances on the canvas do not correspond with equal distances in the real scene. Because the heights of the poles are assumed to be equal, the short pole you drew in the background is much farther away from you than the first pole you drew, despite the fact that they are equally spaced in the drawing.

21. a. The letters with right/left reflection symmetry are A, H, I, M, O, T, U, V, W, X, and Y.
  b. The letters with top/bottom reflection symmetry are B, C, D, E, H, I, K, O, and X.
  c. The letters with both of these symmetries are H, I, O, and X.
  d. The letters with rotational symmetry are H, I, N, O, S, X, and Z.

23. a. Rotating the triangle about its center 120° or 240°, will produce an identical figure.
  b. Rotating the square about its center 90°, 180° or 270° will produce an identical figure.
  c. Rotating the pentagon around its center 72°, 144°, 216°, or 288° will produce an identical figure.
  d. Rotating an n-gon about its center $360°/n$ (and multiples of this angle less than 360°) will produce an identical figure. There are $n-1$ different angles of rotational symmetry for an n-gon.

25. The figure has reflection symmetries (about vertical and horizontal lines drawn through its center), and it has rotation symmetry (an angle of 180°).

27. The figure has 6 reflection symmetries (about 3 diagonal lines drawn through the petals and the center and 3 diagonal lines drawn in the space between the petals and the center), and it has rotation symmetries (angles of 60°, 120°, 180°, 240°, and 300°).

29.

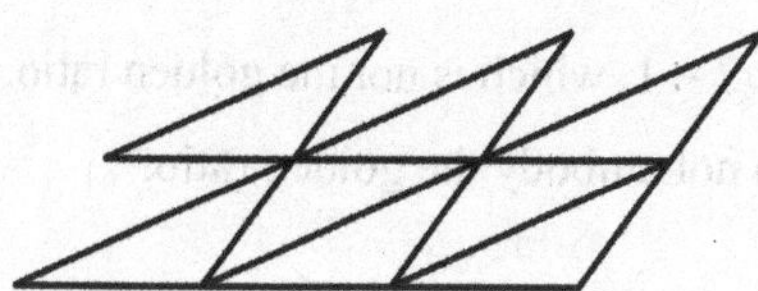

31.

33.

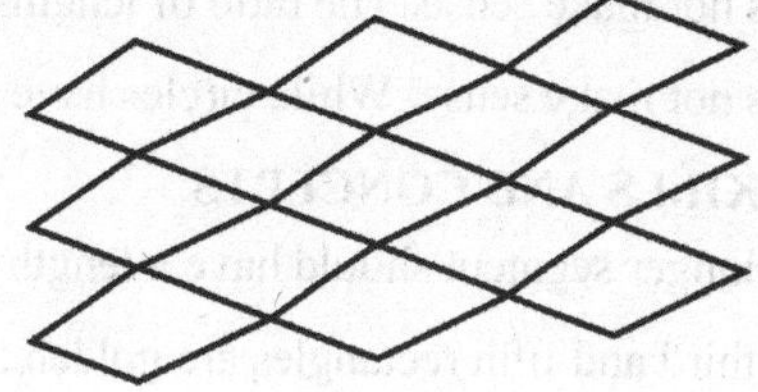

FURTHER APPLICATIONS

35. The angles around a point P are precisely the angles that appear inside of a single quadrilateral. Thus, the angles around P have a sum of 360°, and the quadrilaterals around P fit perfectly together.

UNIT 11C: PROPORTION AND THE GOLDEN RATIO

QUICK QUIZ

1. c. The value of the golden ratio can be derived by solving $\frac{L}{1}=\frac{L+1}{L}$ for L. This equation comes from the idea that the most aesthetically pleasing way to divide a line is one where the ratio of the length of the long piece (L) to the length of the short piece (1) is equal to the ratio of the length of the entire line ($L+1$) to the length of the long piece L, and it is this idea that produces the golden ratio.

2. c. The fourth number in the Fibonacci series is 3, which is not the value for the golden ratio.

3. **c.** The value of ϕ is about 1.6, and the ratio of the long piece to the short piece should be around 1.6 before claiming the line is divided into the golden ratio. Since 0.6/0.4 = 1.5, these are roughly the correct lengths for a golden ratio.

4. **b.** A golden rectangle is defined to be a rectangle whose ratio of its long side to short side is ϕ.

5. **b.** A golden rectangle is defined to be a rectangle whose ratio of its long side to short side is ϕ. Note than $10 \div 6.25 = 1.6$.

6. **a.** The golden rectangle became a cornerstone of their philosophy of aesthetics.

7. **a.** Refer to Figure 11.45.

8. **b.** Because the ratio of successive Fibonacci numbers converges to $\phi \approx 1.62$, each Fibonacci number is about 62% larger than the preceding number.

9. **b.** The 21st number in the series is the sum of the preceding two numbers.

10. **b.** Refer to Table 11.3.

REVIEW QUESTIONS

1. If a line segment is divided in two pieces according to the golden ratio, the ratio of the long piece to the short piece is $\phi = 1.61803\ldots$.

3. The golden ratio and golden rectangle can be found in architecture and many works of art, giving evidence that it is considered visually pleasing.

5. For the Fibonacci sequence, the next number in the sequence is the sum of the previous two numbers, where the first two numbers in the sequence are initially defined.

DOES IT MAKE SENSE?

7. Does not make sense. The ratio of lengths of Maria's two sticks is 2/2 = 1, which is not the golden ratio.

9. Does not make sense. While circles have a lot of symmetry, they do not embody the golden ratio.

BASIC SKILLS AND CONCEPTS

11. The longer segment should have a length of 3.71 inches, and the shorter segment a length of 2.29 inches.

13. The third and fifth rectangles are golden rectangles.

15. If 5.8 m is the longer side, and S the shorter side, we have $5.8 \div S = \phi$, which means $S = 5.8 \div 1.62 = 3.58$ m. If 5.8 m is the shorter side, and L the longer side, we have $L \div 5.8 = \phi$, which means $L = 5.8\phi = 9.40$ m.

17. If 0.66 cm is the longer side, and S the shorter side, we have $0.66 \div S = \phi$, which means $S = 0.66 \div 1.62 = 0.41$ cm. If 0.66 cm is the shorter side, and L the longer side, we have $L \div 0.66 = \phi$, which means $L = 0.66\phi = 1.07$ cm.

FURTHER APPLICATIONS

19. a. Begin with $\frac{L}{1} = \frac{L+1}{L}$, and multiply both sides by L to arrive at $L^2 = L + 1$, which can be rearranged to produce $L^2 - L - 1 = 0$.

b. The two roots are $L = \frac{-(-1) + \sqrt{(-1)^2 - 4(1)(-1)}}{2(1)} = \frac{1+\sqrt{5}}{2}$ and $L = \frac{-(-1) - \sqrt{(-1)^2 - 4(1)(-1)}}{2(1)} = \frac{1-\sqrt{5}}{2}$. The first is $\phi = 1.618034\ldots$.

21. Answers will vary.

23.

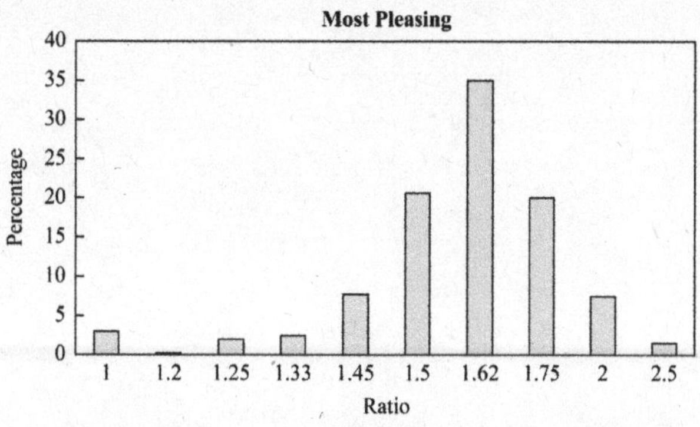

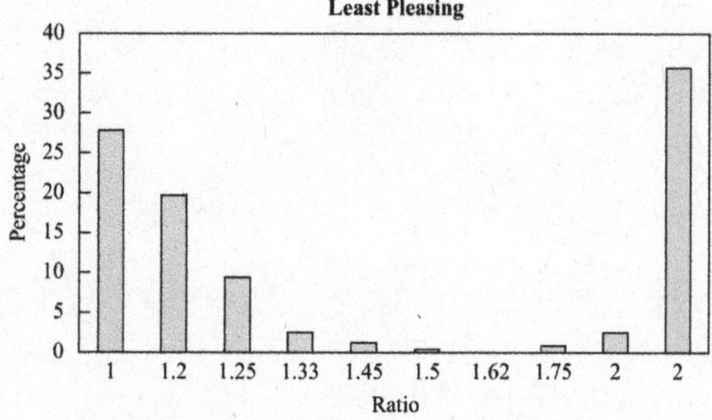

25. a.

a	b	$\frac{a+b}{b}$
38	62	1.6129
28	46	1.6087
56	102	1.5490
56	88	1.6364
24	36	1.6667
77	113	1.6814
40	69	1.5797
46	60	1.7667
15	18	1.8333
39	63	1.6190
53	67	1.7910

b. While the values of $\frac{a+b}{b}$ are clustered around ϕ, their average is 1.6677 and there are more values greater than ϕ. Therefore, the values of $\frac{a+b}{b}$ are not well predicted by ϕ.

c. Answers will vary.

UNIT 12A: VOTING: DOES THE MAJORITY ALWAYS RULE?

QUICK QUIZ

1. **b. In a three-candidate race, it's possible that none of the candidates receives a majority of the votes.**
2. **c. Clinton won 48.3% of the vote, which is not a majority, only a plurality.**
3. **b. A 60% vote is required to end a filibuster, which means 41% of the senators (a minority) can prevent a bill from coming to a vote by staging a filibuster.**
4. **c. In a runoff, the two candidates with the most votes go in a head-to-head competition, and Johnson would be eliminated at this stage.**
5. **b. A preference schedule is a ballot where voters list, in order of their preference, the choices before them.**
6. **c. In an election where preference schedules are used, a voter is expected to fill in all choices.**
7. **a. Look at the second row of Table 12.5.**
8. **c. Candidate D showed up in third place on $12+4+2=18$ ballots.**
9. **b. Candidate E received only 6 first-place votes.**
10. **b. As shown in the text, a different winner would be declared for each of the five methods discussed (plurality, runoff, sequential runoff, a Borda count, and the Condorcet method).**

REVIEW QUESTIONS

1. **Majority rule says that the choice receiving more than 50% of the votes is the winner. A person who wins the majority of votes in an election where only the number of votes is counted is the definitive winner of the election.**

3. **A filibuster occurs when a senator chooses to speak continuously on the senate floor to prevent a vote they oppose from taking place. A filibuster can be ended only by a vote of 3/5, or 60%, of the senators.**

5. **A veto occurs when the President, or any person in an organization with veto power, does not approve a bill or motion to be put into law or practice. In U.S. politics, the veto can be used to keep bills that cannot achieve a 2/3 majority in the house and senate from becoming law, making the bar for enactment higher than a simple majority.**

7. **A preference schedule tells us how many voters chose a particular ranking order among the candidates. See Example 4 for an example of a preference schedule.**

DOES IT MAKES SENSE?

9. **Does not make sense. If both candidates received more than 50% of the vote, the total vote would exceed 100%.**

11. **Makes sense. Imagine a race with three candidates. Herman could win a plurality of the votes in the first election, with Hanna coming in second, in which case a runoff election would ensue. Hanna would win the election if she beat Herman in the runoff.**

13. **Makes sense. This is the way most U.S. presidential elections are decided – a candidate wins both the popular and electoral vote.**

BASIC SKILLS AND CONCEPTS

15. **a. Hayes won $\frac{4,034,142}{8,418,659}=47.9\%$ of the popular vote while Tilden won $\frac{4,286,808}{8,418,659}=50.9\%$ of the popular vote. Tilden won a popular majority.**

    b. Hayes won $\frac{185}{185+184}=50.1\%$ of the electoral vote while Tilden won $\frac{184}{185+184}=49.9\%$ of the electoral vote. Hayes won the electoral vote but did not win the popular vote.

17. a. Harrison won $\frac{5,443,633}{11,388,846} = 47.8\%$ of the popular vote while Cleveland won $\frac{5,538,163}{11,388,846} = 48.6\%$ of the popular vote. Neither candidate received a popular majority.

b. Harrison won $\frac{233}{233+168} = 58.1\%$ of the electoral vote while Cleveland won $\frac{168}{233+168} = 41.9\%$ of the electoral vote. Harrison won the electoral vote but did not win the popular vote.

19. a. Clinton won $\frac{44,909,806}{104,423,923} = 43.0\%$ of the popular vote while Bush won $\frac{39,104,550}{104,423,923} = 37.4\%$ of the popular vote. Neither candidate received a popular majority.

b. Clinton won $\frac{370}{370+168} = 68.8\%$ of the electoral vote while Bush won $\frac{168}{370+168} = 31.2\%$ of the electoral vote. Clinton won the popular vote and the electoral vote.

21. a. Bush won $\frac{50{,}456{,}002}{105{,}405{,}100} = 47.9\%$ of the popular vote while Gore won $\frac{50{,}999{,}897}{105{,}405{,}100} = 48.3\%$ of the popular vote. Neither candidate received a popular majority.

b. Bush won $\frac{271}{271+266} = 50.5\%$ of the electoral vote while Gore won $\frac{266}{271+266} = 49.5\%$ of the electoral vote. Bush won the electoral vote, while Gore won the popular vote.

23. a. Obama won $\frac{69{,}498{,}516}{131{,}313{,}820} = 52.9\%$ of the popular vote while McCain won $\frac{59{,}948{,}323}{131{,}313{,}820} = 45.7\%$ of the popular vote. Obama received a popular majority.

b. Obama won $\frac{365}{365+173} = 67.8\%$ of the electoral vote while McCain won $\frac{173}{365+173} = 32.2\%$ of the electoral vote. Obama won the popular vote and the electoral vote.

25. a. $62/100 = 62\%$ of the senators would vote to end the filibuster, so the filibuster will likely end, and the bill could come to a vote where it would pass.

b. A 2/3 vote of an 11-member jury requires at least 8 votes (because $7/11 \approx 63.6\% < 2/3$ so there will be no conviction in this trial. (A hung jury is likely).

c. Only 72% of the states support the amendment (75% is required), so it will fail to pass.

d. It is not likely to become law because $68/100 = 68\%$ of the Senate will support it, and $270/435 = 62\%$ of the House will support it, and this isn't enough for the veto to be overturned.

27. a. Clinton earned $\frac{44,909,889}{103,756,701} = 43.3\%$ of the popular vote, Bush won $\frac{39,104,545}{103,756,701} = 37.7\%$, and Perot won $\frac{19,742,267}{103,756,701} = 19.0\%$. Clinton won the popular vote by a plurality, but no one won a majority.

b. Clinton won $\frac{370}{538} = 68.8\%$ of the electoral vote, Bush won $\frac{168}{538} = 31.2\%$, and Perot won $\frac{0}{538} = 0\%$. Clinton won both a plurality and a majority of the electoral vote.

c. If Perot had dropped out of the election and Bush had won most of Perot's popular votes, then Bush could have won the popular vote. For Bush to win, his additional popular votes would need to have been distributed among the states in a way that put him ahead of Clinton in many of the states that Clinton had won.

d. If Bush had dropped out of the election and Perot had won most of Bush's popular votes, then Perot could have won the popular vote. However, his additional popular votes would need to have been distributed among states in a way that allowed him to win the electoral vote.

29.

First	A	C	B	C
Second	B	B	C	A
Third	C	A	A	B
Number of Voters	22	20	16	8

31. Use the preference table from Exercise 27.

a. A has 22 votes, B has 16 votes, and C has 28 votes, so C wins.

b. The single runoff would be between A and C, with A having 22 votes and C having 44 votes, so C wins.

c. B is eliminated first, which leaves a runoff between A and C, which is the same as part (b), so C wins.

d. The count for candidate A is $22\times3+20\times1+16\times1+8\times2=118$.

The count for candidate B is $22\times2+20\times2+16\times3+8\times1=140$.

The count for candidate C is $22\times1+20\times3+16\times2+8\times3=138$.

So B wins.

e. A vs. B: B wins 36 to 30.

A vs. C: C wins 44 to 22.

B vs. C: B wins 38 to 28.

B wins 2 out of 3 pairwise comparisons, so B wins.

33. 22 preferred B to E.

35. With E removed and looking at the first row of the preference table: A would receive 18, B would receive 16, C would receive 12, and D would receive 9.

37. a. A total of 66 votes were cast.

b. Candidate D is the plurality winner (22 first-place votes), but not by a majority.

c. In a runoff, candidate D would win, because 40 voters prefer D to B (who finished in second place in part (b)), while only 26 prefer B to D.

d. In a sequential runoff, the candidate with the fewest votes is eliminated at each stage. Candidate A received the fewest first-place votes in the original results (only 8), so A is eliminated, and those votes are redistributed among the other candidates. Recounting the votes after A has been eliminated, we find that B has 20 first-place votes, C has 16, and D has 30. Thus C is now eliminated, and the votes counted again. In this final stage, we find that B has 26 first-place votes, and D has 40 votes (as in part (b)), so D wins the sequential runoff.

e. In a Borda count, first-place votes receive a value of 4, second-place votes receive a value of 3, and so on. Working through each column of the table, and noting the number of voters who cast such ballots, the count for candidate A is computed as $20\times1+15\times3+10\times2+8\times4+7\times3+6\times3=156$.

The count for candidate B is $20\times4+15\times1+10\times1+8\times1+7\times2+6\times2=139$.

The count for candidate C is $20\times2+15\times2+10\times4+8\times2+7\times1+6\times4=157$.

The count for candidate D is $20\times3+15\times4+10\times3+8\times3+7\times4+6\times1=208$.

Candidate D wins in this case.

f. Since there are 4 candidates, there are 6 pairings to consider. The results for each pairing are shown below.

A beats B, 46 to 20.
C beats B, 39 to 27.
C beats A, 36 to 30.
D beats B, 40 to 26.
D beats A, 52 to 14.
D beats C, 50 to 16.

Since D wins more pairwise comparisons than any other candidate, D is the winner.

g. As the winner by all five methods, candidate D is clearly the winner of the election.

39. a. A total of 100 votes were cast.

b. Candidate C is the plurality winner (40 first-place votes), but not by a majority.

c. In a runoff, candidate A would win, because 55 voters prefer A to C.

d. With only three candidates, the sequential runoff method is the same as the standard runoff between the top two candidates, and thus A would win.

e. Using the process illustrated in Exercise 35e, the counts of the three candidates are $A = 200$, $B = 210$, and $C = 190$. Candidate B has the highest count, so B wins a Borda count.

f. Since there are 3 candidates, there are 3 pairings to consider. The results of each pairing are:

B beats A, 55 to 45. A beats C, 55 to 45. B beats C, 55 to 45.

Since B wins two of the pairwise comparisons, B is the winner.

g. Candidate A wins by runoff, B wins by a Borda count and pairwise comparisons, and C is the plurality winner. Thus there is no clear winner.

41. a. A total of 90 votes were cast.

b. Candidate E is the plurality winner (40 first-place votes), but not by a majority.

c. Candidate B wins a runoff, because 50 voters prefer B over E, who would receive 40 votes (D is eliminated at the beginning).

d. With only three candidates earning first-place votes, the sequential runoff method is the same as the standard runoff between the top two candidates, and thus B would win.

e. Using the process illustrated in Exercise 35e, the counts of the candidates are $A = 260$, $B = 290$, $C = 200$, $D = 290$, and $E = 310$. Candidate E has the highest count, so E wins a Borda cont.

f. Since there are 5 candidates, there are 10 pairings to consider. The results for each pairing are shown below.

A beats B, 60 to 30.
A beats C, 60 to 30.
D beats A, 60 to 30.
E beats A, 70 to 20.
B beats C, 90 to 0.
D beats B, 60 to 30.
B beats E, 50 to 40.
D beats C, 60 to 30.
C beats E, 50 to 40.
E beats D, 70 to 20.

Since D wins more pairwise comparisons than any other candidate (3), D is the winner.

g. Candidates B and E each win two of the five methods, and D wins one method, so all have some claim on the prize. The outcome is debatable.

FURTHER APPLICATIONS

43. a. Best won a plurality, because he won more votes than the other candidates, but no one won a majority, as more than 50% of the vote is required before a majority winner can be claimed.

b. Able needs just over 15 percentage points to gain a majority and win a runoff. Since $15\%/23\% = 0.652$, Able needs just over 65.2% of Crown's votes.

45. a. Irving won a plurality because he won more votes than the other candidates, but no one won a majority (Irving won $205/485 = 42.3\%$ of the vote, which isn't enough).

b. To overtake Irving, Heyduke needs 83 additional votes (this would give Heyduke a total of 243 votes, with Irving winning only 242 votes).

47. Since there are 4 candidates, there are 6 pairings to consider. The results for each pairing are shown below.

B wins over A, 60 to 50.
A wins over C, 110 to 0.
A wins over D, 110 to 0.
B wins over C, 60 to 50.
B wins over D,60 to 50.

At this point, we don't even need to check the last pairing because we have already found the solution: candidate B wins all three of his head-to-head races, and thus is the Condorcet winner. (If you are curious, C beats D, 60 votes to 50 votes).

49. a. Label the four candidates as A, B, C, and D. It is easy enough to list all possible pairings: AB, AC, AD, BC, BD, and CD, for a total of 6 pairings. Note that this list is the list of all combinations of the four objects A, B, C, and D taken two at a time, so there should be ${}_4C_2 = 6$ pairings.

b. With 5 candidates, there are ${}_5C_2 = 10$ pairings.

c. With 6 candidates, there are ${}_6C_2 = 15$ pairings.

51. Each voter will record a first, second, third, and fourth place vote. The points for these places total $4+3+2+1=10$ points. With 30 voters, there are $30\times 10 = 300$ total points. Candidate D has $300-100-80-75=45$ points, and thus candidate A wins the election.

UNIT 12B: THEORY OF VOTING

QUICK QUIZ

1. c. An election isn't declared fair until all four fairness criteria are satisfied.
2. a. Berman wins a plurality because he has the most votes.
3. b. Goldsmith won the least number of first-place votes, and so is eliminated in a runoff. Freedman wins Goldsmith's 18 votes, and so beats Berman.
4. c. Criterion 1 is only applicable when a candidate wins a majority of votes.
5. b. Freedman wins both head-to-head races, and thus by Criterion 2, Freedman should win if this is a fair election.
6. a. See Exercise 5.
7. a. Candidate X (Freedman) needs to be declared the winner of the first election before Criterion 4 comes into play.
8. b. There are plenty of elections where all four fairness criteria are satisfied. Arrow's theorem says only that no election system can be satisfied in *all* circumstances.
9. b. The text gives examples of approval voting scenarios that are not fair (and Arrow's theorem guarantees there are such cases).
10. c. In a small state, the number of voters per senator is much smaller than in a large state, and thus each voter has more voting power.

REVIEW QUESTIONS

1. See the Information Box on page 675.

3. Approval voting, asks voters to specify whether they approve or disapprove of each candidate. Voters may approve as many candidates as they like, and the candidate with the most approval votes wins.

DOES IT MAKE SENSE?

5. Makes sense. Karen feels that because she won all head-to-head races, she should win the election so that Criterion 2 is satisfied – otherwise, it's not a fair election.

7. Does not make sense. Assuming the plurality method is used to decide the second election, Table 12.9 claims Criterion 3 is always satisfied (and thus Wendy could not have managed to win).

BASIC SKILLS AND CONCEPTS

9. A candidate who wins a majority is the only person to receive a plurality. Thus the candidate will win the election, and Criterion 1 is satisfied.

11. The following preference schedule is just one possible example.

First	B	A	C	C
Second	A	B	A	B
Third	C	C	B	A
Number of Voters	2	4	2	3

C is the plurality winner, but A beats both C and B, which means Criterion 2 is violated.

13. Candidate A would win by the plurality method. However if candidate C were to drop out of the election, then B would win by the plurality method, and Criterion 4 would be violated.

15. Assume a candidate receives a majority. In either runoff method, votes are redistributed as candidates are eliminated. But it is impossible for another candidate to accumulate enough votes to overtake a candidate who already has a majority.

17. By the sequential runoff method (which is the same as the top two runoff method with three candidates), candidate B is eliminated, and C wins the runoff. However, candidate B wins head-to-head races against A and C, so Criterion 2 is violated.

19. In the sequential runoff method, candidate B is eliminated first, then candidate A, making C the winner. Now suppose that the 4 voters on the third ballot (ACB) move C up and vote for the ranking CAB. Now A is eliminated first, and B wins the election. Thus criterion 3 is violated.

21. The following preference schedule is just one possible example.

First	A	B	C
Second	B	A	B
Third	C	C	A
Number of Voters	4	3	5

After candidate B is eliminated, A wins the runoff. However, if C were to drop out, B would win the election. So Criterion 4 is violated.

23. The following preference schedule is just one possible example.

First	A	B	C
Second	B	C	B
Third	C	A	A
Number of Voters	4	2	1

Candidate A has a majority of the votes, but loses in a Borda count (A gets 8 points, B gets 9, and C gets 4, using 2, 1, and 0 points).

25. The following preference schedule is just one possible example.

First	A	D	C
Second	B	B	B
Third	C	A	D
Fourth	D	C	A
Number of Voters	4	8	3

Using the Borda count (with 3, 2, 1, and 0 points), A gets 20 points, B gets 30, C gets 13, and D gets 27 points, making B the winner. However, D wins head-to-head races against all other candidates, and thus Criterion 2 is violated.

27. Using the point system (with 2, 1, and 0 points), candidate A gets 11 points, B gets 11 points, and C gets 14 points, which makes C the winner. However, if Candidate A were to drop from the race, then B would receive 7 points and C would receive 5 points, making B the winner (notice that the point values become 1 and 0 with only two candidates). Thus Criterion 4 is violated.

29. Assume candidate A wins a majority of first-place votes. Then in every head-to-head race with another candidate, A must win (by a majority). Thus, A wins every head-to-head race and is the winner by pairwise comparisons, so Criterion 1 is satisfied.

31. Suppose candidate A wins by the method of pairwise comparisons, and in a second election, moves up above candidate B in at least one ballot. A's position relative to B remains the same or improves, and A's position and B's position relative to the other candidates remains the same, so A must win the second election.

33. The following preference schedule is just one possible example.

First	A	A	B	E	E
Second	B	C	A	B	D
Third	C	D	C	A	B
Fourth	D	E	D	C	A
Fifth	E	B	E	D	C
Number of Voters	1	3	2	1	2

By the pairwise comparison method, A would beat C, D, and E; B would beat A and C; C would beat D and E; D would beat B and E; and E would beat B. This would lead to A winning. However, if E were to drop out of the election, then A and B would have two pairwise wins, and C and D would have one pairwise win so there is no winner by this method. Criterion 4 is violated because the outcome of the election is changed when E drops out.

35. a. Voting only for their first choices, candidate C wins by plurality with 42% of the vote.

b. By an approval vote, $28\% + 29\% = 57\%$ of the voters approve of A, $28\% + 29\% + 1\% = 58\%$ approve of B, and 42% of the voters approve of C. The winner is B.

37. The electoral votes per person for each of the states can be computed by dividing the number of electoral votes by the population. The results are as follows:

New York: $\frac{29}{19{,}745{,}000} = 1.47 \times 10^{-6}$ Rhode Island: $\frac{4}{1{,}056{,}000} = 3.79 \times 10^{-6}$

Illinois: $\frac{20}{12{,}802{,}000} = 1.56 \times 10^{-6}$ Alaska: $\frac{3}{742{,}000} = 4.04 \times 10^{-6}$

It is evident that voters in Alaska have more voting power than those in Illinois.

39. The voters in Rhode Island have more voting power than those in Illinois (see Exercise 37).

41. From the greatest to least voting power, per person, the ranking is: Alaska, Rhode Island, Illinois, New York

FURTHER APPLICATIONS

43. A single runoff would eliminate C and D. Candidate A would win the runoff, 23 to 18. There is no majority winner, so Criterion 1 does not apply. Candidate A beats all other candidates one-on-one, and is also declared winner in a single runoff, so Criterion 2 is satisfied. If A is moved up in any of the rankings, it doesn't affect the outcome of the election, so Criterion 3 is satisfied. If any combination of {B, C, D} drops out of the race, the outcome is not changed. So in this case (though not in general), Criterion 4 is satisfied.

45. Candidate A wins the point system (using 3, 2, 1, and 0 points) with 83 points. There is no majority winner, so Criterion 1 does not apply. Candidate A beats all other candidates one-on-one, and is also declared winner by the point system, so Criterion 2 is satisfied. The point system always satisfies Criterion 3. If any combination of {B, C, D} drops out of the race, the outcome is not changed. So in this case (though not in general), Criterion 4 is satisfied.

47. Candidate A wins by a plurality, but not by a majority. There is no majority winner, so Criterion 1 does not apply. Candidate E beats all other candidates one-on-one, but loses by the plurality method, so Criterion 2 is violated. The plurality method always satisfies Criterion 3. If any of B, C, or D were to drop out of the election, the outcome of A winning would change, so Criterion 4 is violated.

49. The candidates E, D, and B are eliminated sequentially, leaving a final runoff between A and C, which C wins. There is no majority winner, so Criterion 1 does not apply. Candidate E beats all other candidates one-on-one, but loses in the sequential runoff method, so Criterion 2 is violated. If candidate C moved up in any of the rankings, the outcome is not affected, so Criterion 3 is satisfied. If A were to drop out of the election, D would win instead of C, so Criterion 4 is violated.

51. We have seen that E beats all other candidates in pairwise races (see Exercise 47) and is the winner by the pairwise comparison method. There is no majority winner, so Criterion 1 does not apply. The pairwise comparison method always satisfies Criteria 2 and 3. It can be shown that if any combination of {A, B, C, D} drops out of the race, the winner is still E, so Criterion 4 is satisfied.

UNIT 12C: APPORTIONMENT: THE HOUSE OF REPRESENTATIVES AND BEYOND

QUICK QUIZ

1. **c.** The Constitution does not specify the number of representatives, and the House has, in fact, had 435 members only since 1912. The number briefly rose by two when Hawaii and Alaska were granted statehood (1959), but dropped back to 435 in the next apportionment (and will likely remain at that level, unless a 1941 law is changed).

2. **b.** Apportionment is a process used to divide the available seats among the states.

3. **a.** The standard divisor is defined to be the total U.S. population divided by the number of seats in the House.

4. **b.** The standard quota is defined to be the population of a state divided by the standard divisor. In this case, the standard divisor is 1 million.

5. **b.** In this scenario, the standard divisor would be defined as the population of students divided by the number of teachers, or $25,000/1000 = 25$.

6. **b.** In this scenario, the standard quota would be defined as the population of the school divided by the standard divisor computed in Exercise 5. Thus it would be $220/25 = 8.8$.

7. **c.** Parks Elementary would get the eight teachers because the Hamilton method assigns the extra teacher to the school with the highest standard quota.

8. **c.** The Hamilton, Jefferson, Webster, and Hill-Huntington methods are the only four methods of apportionment that have been used (to date) to assign the seats in the House to various states.

9. **c.** At present, the law states that the Hill-Huntington method is to be used to reapportion seats at every census.

10. **b.** As shown in the text, the methods currently used do not always satisfy the quota criterion.

REVIEW QUESTIONS

1. For the U.S. House of Representatives, apportionment is a process used to divide the available seats among the states. More generally, apportionment is a process used to divide a set of people or objects among various groups or individuals.

3. See the Information Box on page 689 for a description of Hamilton's method. The use of Hamilton's method was vetoed in 1792. Hamilton's method was then adopted in 1850, but abandoned in 1900.

5. See the Information Box on page 693 for a description of Jefferson's method. Jefferson's method was signed into law in 1792 and used through the 1830s. Congress abandoned Jefferson's method for apportionment following the 1840 census.

7. Unlike Jefferson's method, Webster's method seeks modified quotas that give the correct total number of seats by using standard rounding rules. In the Hill-Huntington method, rounding is based instead on the geometric mean of the integers on either side of the modified quota.

DOES IT MAKE SENSE?

9. Makes sense. If the number of staff support persons needed in a division depends on the number of employees in that division, an apportionment method would be a good idea.

11. Does not make sense. All apportionment methods have deficiencies, and no single method (of those discussed in the text) is better than the others due to the level of math required to carry it out.

BASIC SKILLS AND CONCEPTS

13. The number of people per representative would be $\frac{350,000,000}{435} = 804,598.$

If the constitutional limit were observed, the number of representatives would be $\frac{350,000,000}{30,000} = 11,667.$

15. In order to compute the standard quota, the standard divisor must first be computed. The standard divisor is $\frac{309 \text{ million}}{435} = 710,344.8.$ For Connecticut, the standard quota is $\frac{3,574,097}{710,344.8} = 5.03,$ which is close to the 5 seats the state actually has, so Connecticut is very slightly underrepresented.

17. The standard divisor is 710,345 (see Exercise 15). For Florida, the standard quota is $\frac{18,801,310}{710,344.8} = 26.47,$ which is smaller than the 27 seats the state actually has, so Florida is overrepresented.

19. The total number of employees is $250+320+380+400=1350.$ The standard divisor is then $\frac{1350}{35} = 38.57.$

For the first division, the standard quota is $\frac{250}{38.57} = 6.48.$ The standard quotas for the other divisions are 8.30, 9.85, and 10.37, respectively (computed in a similar fashion).

21. The standard divisor is $\frac{5000}{100} = 50.$

STATE	A	B	C	D	Total
Population	914	1186	2192	708	5000
Standard Quota	$\frac{914}{50} = 18.28$	$\frac{1186}{50} = 23.72$	$\frac{2192}{50} = 43.84$	$\frac{708}{50} = 14.16$	100
Minimum Quota	18	23	43	14	98
Fractional Remainder	0.28	0.72	0.84	0.16	N/A
Final Apportionment	18	24	44	14	100

23. Refer to Exercise 19. Using Hamilton's method for the assignments, round each standard quota down to get $6+8+9+10=33$ technicians. Since the first and third divisions have the highest remainders, the extra two technicians will be assigned to them, giving a final apportionment of 7, 8, 10, and 10 technicians to each of the four divisions, respectively.

25. The total population is $950+670+246=1866$, so with 100 seats to be apportioned, the standard divisor is $1866/100=18.66$. This is used to compute the standard quota and the minimum quotas in the table below. Hamilton's method applied to these three states yields:

State	A	B	C	Total
Population	950	670	246	1866
Standard Quota	50.91	35.91	13.18	100
Minimum Quota	50	35	13	98
Fractional Remainder	0.91	0.91	0.18	2
Final Apportionment	51	36	13	100

Assuming 101 delegates, Hamilton's method yields:

State	A	B	C	Total
Population	950	670	246	1866
Standard Quota	50.42	35.26	13.32	101
Minimum Quota	51	36	13	100
Fractional Remainder	0.42	0.26	0.32	1
Final Apportionment	52	36	13	101

No state lost seats as a result of the additional available seat, so the Alabama paradox does not occur here.

27. The total population is $770+155+70+673=1668$, so with 100 seats to be apportioned, the standard divisor is $1668/100=16.68$. This is used to compute the standard quota and the minimum quotas in the table below. Hamilton's method applied to these four states yields:

State	A	B	C	D	Total
Population	770	155	70	673	1668
Standard Quota	46.16	9.29	4.20	40.35	100
Minimum Quota	46	9	4	40	99
Fractional Remainder	0.16	0.29	0.20	0.35	1
Final Apportionment	46	9	4	41	100

Assuming 101 delegates, Hamilton's method yields:

State	A	B	C	D	Total
Population	770	155	70	673	1668
Standard Quota	46.62	9.39	4.24	40.75	101
Minimum Quota	46	9	4	40	99
Fractional Remainder	0.62	0.39	0.24	0.75	2
Final Apportionment	47	9	4	41	101

No state lost seats as a result of the additional available seat, so the Alabama paradox does not occur here.

29. The total population is $98+689+212=999$, so with 100 seats to be apportioned, the standard divisor is $999/100=9.99$. This is used to compute the standard quota and the minimum quotas in the table below. Using a modified divisor of 9.83 instead, we get the modified quotas listed, and the new minimum quotas. Jefferson's method then yields:

State	A	B	C	Total
Population	98	689	212	999
Standard Quota	9.81	68.97	21.22	100
Minimum Quota	9	68	21	98
Modified Quota	9.97	70.09	21.57	101.63
New Minimum Quota	9	70	21	100

Since the new minimum quota successfully apportions all 100 seats, we can stop. Note, however, that the quota criterion is violated, because state B's standard quota is 68.97, yet it was given 70 seats.

31. The total population is 979, so with 100 seats to be apportioned, the standard divisor is $979/100=9.79$. This is used to compute the standard quota and the minimum quotas in the table below. Using a modified divisor of 9.60 instead, we get the modified quotas listed, and the new minimum quotas. Jefferson's method then yields:

State	A	B	C	D	Total
Population	69	680	155	75	979
Standard Quota	7.05	69.46	15.83	7.66	100
Minimum Quota	7	69	15	7	98
Modified Quota	7.19	70.83	16.15	7.81	101.98
New Minimum Quota	7	70	16	7	100

Since the new minimum quota successfully apportions all 100 seats, we can stop. The quota criterion is satisfied.

33. After trial and error, a modified divisor of 38.4 was found to work. The results are summarized in the table below.

Division	I	II	III	IV
Number in division	250	320	380	400
Modified Quota	6.51	8.33	9.90	10.42
Number assigned	7	8	10	10

35. An interesting situation arises in this problem: if the standard divisor, standard quotas, and geometric means are computed without rounding, it turns out a modified divisor is not necessary. That is, after computing said values, the Hill-Huntington method immediately produces an apportionment with no leftover technicians, and it is not necessary to seek a modified divisor. The results are summarized in the table below.

Division	I	II	III	IV
Number in division	250	320	380	400
Standard Quota	6.4815	8.30	9.85	10.37
Geometric Mean	$\sqrt{6\times7}=6.48$	$\sqrt{8\times9}=8.49$	$\sqrt{9\times10}=9.49$	$\sqrt{10\times11}=10.49$
Number assigned	7	8	10	10

35. (continued)

Division I is the culprit – the standard quota and geometric mean are so close to one another that it is hard to determine whether one should round up or down (note that the values at the other entries in the table are not shown to the same precision, because the decision about whether to round up or down can be made with less precise values). If the rounded values from Exercise 19 are used, you will need to use a modified divisor in order to find an apportionment that uses all 35 technicians. A divisor of 38.4 is suitable (and realize that modified divisors are not unique, in that divisors close to one another produce the same results).

FURTHER APPLICATIONS

37. The total population is 7710, so with 100 seats to be apportioned, the standard divisor is $\frac{7710}{100} = 77.10$. This is used to compute the standard quota and the minimum quotas in the table below. Hamilton's method applied to these three states yields:

State	A	B	C	Total
Population	1140	6320	250	7710
Standard Quota	14.79	81.97	3.24	100
Minimum Quota	14	81	3	98
Fractional Remainder	0.79	0.97	0.24	2
Final Apportionment	15	82	3	100

With the addition of a new state D of population 500, for whom 5 new delegates are added, Hamilton's method, with a new standard divisor of $\frac{8210}{105} = 78.19$, yields:

State	A	B	C	D	Total
Population	1140	6320	250	500	8210
Standard Quota	14.58	80.83	3.20	6.39	105
Minimum Quota	14	80	3	6	103
Fractional Remainder	0.58	0.83	0.20	0.39	2
Final Apportionment	15	81	3	6	105

Even though 5 new seats were added with state D representation in mind, Hamilton's method assigned 6 seats to this state, 1 of them at the expense of state B. Since B lost a seat as a result of the additional seats for the new state, the New State paradox occurs here.

39. a. The total population is 999, so with 100 seats to be apportioned, the standard divisor is $\frac{999}{100} = 99.9$. This is used to compute the standard quota and the minimum quotas in the table below. Hamilton's method yields:

State	A	B	C	Total
Population	535	334	120	999
Standard Quota	53.55	34.43	12.01	100
Minimum Quota	53	34	12	99
Fractional Remainder	0.55	0.43	0.01	1
Final Apportionment	54	34	12	100

39. (continued)

b. Jefferson's method begins as with Hamilton's. As noted in part (a), the standard divisor is 9.99, so we try lower modified divisors until the apportionment comes out just right. By trial and error, the modified divisor 9.90 is found to work, as documented in the last two rows of the table below. Note that this choice of the modified divisor is not unique, as other nearby values also work.

State	A	B	C	Total
Population	535	334	120	999
Standard Quota	53.55	34.43	12.01	100
Minimum Quota	53	34	12	98
Modified Quota	54.04	34.75	12.12	100.91
New Minimum Quota	54	34	12	100

c. Webster's method requires us to find a modified divisor such that the corresponding modified quotas round (not truncate) to numbers that sum to the desired 100 delegates. Inspecting the table in part (a), we see that the standard divisor and standard quotas are already adequate – we don't need to seek a modified divisor in this case.

State	A	B	C	Total
Population	535	334	120	999
Standard Quota	53.55	34.43	12.01	100
Minimum Quota	53	34	12	99
Rounded Quota	54	34	12	100

d. The Hill-Huntington method requires us to find a modified divisor such that the corresponding modified quotas, *rounded relative to the geometric mean*, yield numbers which sum to the required 100 delegates. As in part (c), the standard divisor and standard quotas are already sufficient because they round (per the geometric means) to the desired apportionment. Thus there is no need to seek out a modified divisor in this case.

State	A	B	C	Total
Population	535	334	120	999
Standard Quota	53.55	34.43	12.01	100
Minimum Quota	53	34	12	99
Geometric Mean	53.50	34.50	12.49	
Rounded Quota	54	34	12	100

The geometric means are for the whole numbers bracketing the standard quotas, namely: $\sqrt{53\times54}=53.50$, $\sqrt{34\times35}=34.50$, and $\sqrt{12\times13}=12.49$. The modified quotas (in this case the standard quotas) are compared to these, and hence 34.43 and 12.01 are rounded down to 34 and 12, respectively, whereas 53.55 is rounded up to 54.

e. All four methods gave the same results.

41. Refer to Exercise 39 for a detailed explanation of the method of solution that yields the following apportionments.

a. Standard divisor = 100

State	A	B	C	D	Total
Population	836	2703	2626	3835	10,000
Standard Quota	8.36	27.03	26.26	38.35	100
Minimum Quota	8	27	26	38	99
Fractional Remainder	0.36	0.03	0.26	0.35	1
Final Apportionment	9	27	26	38	100

41. (continued)

b. Modified divisor = 98.3

State	A	B	C	D	Total
Population	836	2703	2626	3835	10,000
Standard Quota	8.36	27.03	26.26	38.35	100
Minimum Quota	8	27	26	38	99
Modified Quota	8.50	27.50	26.71	39.01	101.73
New Minimum Quota	8	27	26	39	100

c. Modified divisor = 99.5

State	A	B	C	D	Total
Population	836	2703	2626	3835	10,000
Standard Quota	8.36	27.03	26.26	38.35	100
Minimum Quota	8	27	26	38	99
Modified Quota	8.40	27.17	26.39	38.54	100.50
Rounded Quota	8	27	26	39	100

d. Modified divisor = 99.5.

State	A	B	C	D	Total
Population	836	2703	2626	3835	10,000
Standard Quota	8.36	27.03	26.26	38.35	100
Minimum Quota	8	27	26	38	99
Modified Quota	8.40	27.17	26.39	38.54	100.49
Geometric Mean	8.49	27.50	26.50	38.50	-
Rounded Quota	8	27	26	39	100

e. The Jefferson, Webster, and Hill-Huntington methods all gave the same result, so one could argue that these yield the best apportionment.

43. a. The total population is 390 students, so with 10 committee positions to be apportioned, the standard divisor is $\frac{390}{10} = 39$. This is used to compute the standard quota and the minimum quotas in the table below.

Hamilton's method yields:

Group	Social	Political	Athletic	Total
Population	48	97	245	390
Standard Quota	1.23	2.49	6.28	10
Minimum Quota	1	2	6	9
Fractional Remainder	0.23	0.49	0.28	1
Final Apportionment	1	3	6	10

b. Jefferson's method begins as with Hamilton's. As noted in part (a), the standard divisor is 39, so we try lower modified divisors until the apportionment comes out just right. By trial and error, the modified divisor of 35 is found to work. Note that this choice of the modified divisor is not unique, as other nearby values also work.

Group	Social	Political	Athletic	Total
Population	48	97	245	390
Standard Quota	1.23	2.49	6.28	10
Minimum Quota	1	2	6	9
Modified Quota	1.37	2.77	7.00	11.14
New Minimum Quota	1	2	7	10

43. (continued)

c. Webster's method requires us to find a modified divisor such that the corresponding modified quotas round (not truncate) to numbers that sum to the desired 10 members. Inspecting the tables in parts **a** and **b**, we see that neither the standard nor modified quotas there work, so we must try other modified divisors. This time, we find that 38 is a suitable modified divisor. Note that this choice of the modified divisor is not unique, as other nearby values also work.

Group	Social	Political	Athletic	Total
Population	48	97	245	390
Standard Quota	1.23	2.49	6.28	10
Minimum Quota	1	2	6	9
Modified Quota	1.26	2.55	6.45	10.26
Rounded Quota	1	3	6	10

d. The Hill-Huntington method requires us to find a modified divisor such that the corresponding modified quotas, *rounded relative to the geometric mean*, yield numbers that sum to the desired 10 members. The standard divisor of 39 and standard quotas are already sufficient because they round (per the geometric means) to the desired apportionment. Thus there is no need to seek out a modified divisor in this case.

Group	Social	Political	Athletic	Total
Population	48	97	245	390
Standard Quota	1.23	2.49	6.28	10
Minimum Quota	1	2	6	9
Geometric Mean	1.41	2.45	6.48	-
Rounded Quota	1	3	6	10

The geometric means are for the whole numbers bracketing the standard quotas, namely: $\sqrt{1\times2}=1.41$, $\sqrt{2\times3}=2.45$, and $\sqrt{6\times7}=6.48$. The modified quotas (in this case the standard quotas) are compared to these, and hence 1.23 and 6.28 are rounded down to 1 and 6, respectively, whereas 2.49 is rounded up to 3.

e. The Hamilton, Webster, and Hill-Huntington methods all give the same result, and thus one could argue that each of these yield the best apportionment.

45. Refer to Exercise 43 for a detailed explanation of the method of solution that yields the following apportionments.

a. Note that the total "population" is $2.5+7.6+3.9+5.5=19.5$ (million dollars).

Standard divisor = 0.78

Store	Boulder	Denver	Broomfield	Ft. Collins	Total
Population	2.5	7.6	3.9	5.5	19.5
Standard Quota	3.21	9.74	5.00	7.05	25
Minimum Quota	3	9	5	7	24
Fractional Remainder	0.21	0.74	0.00	0.05	1
Final Apportionment	3	10	5	7	25

b. Modified divisor = 0.76

Store	Boulder	Denver	Broomfield	Ft. Collins	Total
Population	2.5	7.6	3.9	5.5	19.5
Standard Quota	3.21	9.74	5.00	7.05	25
Minimum Quota	3	9	5	7	24
Modified Quota	3.29	10.00	5.13	7.24	25.66
New Minimum Quota	3	10	5	7	25

45. (continued)

c. Modified divisor not necessary.

Store	Boulder	Denver	Broomfield	Ft. Collins	Total
Population	2.5	7.6	3.9	5.5	19.5
Standard Quota	3.21	9.74	5.00	7.05	25
Minimum Quota	3	9	5	7	24
Rounded Quota	3	10	5	7	25

d. Modified divisor not necessary.

Store	Boulder	Denver	Broomfield	Ft. Collins	Total
Population	2.5	7.6	3.9	5.5	19.5
Standard Quota	3.21	9.74	5.00	7.05	25
Minimum Quota	3	9	5	7	24
Geometric Mean	3.46	9.49	5.48	7.48	
Rounded Quota	3	10	5	7	25

e. All four methods give the same results.

UNIT 12D: DIVIDING THE POLITICAL PIE

QUICK QUIZ

1. **c.** The process of redrawing district boundaries is called redistricting.
2. **b.** This answer does the best job in summarizing the political importance of redistricting.
3. **b.** As shown in Table 12.19 in the text, house elections are decided, on average, by larger margins of victory.
4. **c.** If district boundaries were drawn in a random fashion, one would expect that Republicans would win about 54% of the house seats. Since they won only 6/13 = 46% of the seats, it appears that the district boundaries were set in a way that favored Democrats.
5. **b.** If district boundaries were drawn in a random fashion, one would expect that Democrats would win about 51% of the house seats. Since they won 4/13 = 31% of the seats, it appears that the district boundaries were set in a way that favored Republicans.
6. **c.** Gerrymandering is the drawing of district boundaries so as to serve the political interests of the politicians in charge of the drawing process.
7. **a.** If you concentrate most of the Republicans in a few districts (and essentially concede that they will win in those districts), you will stack the deck in the favor of the Democrats in all the remaining districts.
8. **b.** The courts have generally allowed even very convoluted district boundaries to stand as long as they don't violate existing laws.
9. **a.** If district lines are drawn in such a way as to maximize the number of, say, Democratic seats in the House, they must concentrate the Republican voters in a few districts, and these districts are going to elect Republican representatives (assuming voters vote along party lines).
10. **b.** In an election for a seat that, say, a Democrat is almost guaranteed to win, the real contest occurs in the primary election, rather than the general election. Primaries tend to draw smaller numbers of voters, and those with more clearly partisan interests, and this results in the election of a representative with more extreme partisan views.

REVIEW QUESTIONS

1. Redistricting is the redrawing of district boundaries that each member of the House of Representatives represent to reflect shifts in population. It occurs every 10 years, after a census.

3. Gerrymandering is the drawing of district boundaries so as to serve the political interests of the politicians in charge of the drawing process. The term originated in 1812, when Massachusetts Governor Elbridge Gerry created a district that critics ridiculed as having the shape of a salamander.

5. All districts within a particular state must have very nearly equal populations and each district must be contiguous, meaning that every part of the district must be connected to every other part.

DOES IT MAKE SENSE?

7. Makes sense. The district lines in that state are most likely drawn in such a way as to favor Democrats.
9. Does not make sense. If the practice of gerrymandering is alive and well in your state, you probably shouldn't expect that the percentage of voters from the various political parties is accurately reflected in the percentage of representatives from those parties that win elections.
11. Does not make sense. Current laws require that district lines should be drawn to produce contiguous districts.

BASIC SKILLS AND CONCEPTS

13. a. In 2010, the percentage of votes cast for Republicans was $\frac{2053}{2053+1611} = 56\%$, and for Democrats it was $\frac{1611}{2053+1611} = 44\%$. In 2012, the percentage of votes cast for Republicans was $\frac{2620}{2620+2412} = 52\%$, and for Democrats it was $\frac{2412}{2620+2412} = 48\%$.

b. In 2010, the percentage of House seats won by Republicans was $\frac{13}{18} = 72\%$, and for Democrats it was $\frac{5}{18} = 28\%$. In 2012, the percentage of House seats won by Republicans was $\frac{12}{16} = 75\%$, and it was $\frac{4}{16} = 25\%$ for Democrats.

c. In both 2010 and 2012, the percentage of House seats won by Republicans was significantly larger than the percentage of votes they received.

d. Redistricting does not seem to have had an effect on the distribution of representatives.

15. a. In 2010, the percentage of votes cast for Republicans was $\frac{3058}{3058+1450} = 68\%$, and for Democrats it was $\frac{1450}{3058+1450} = 32\%$. In 2012, the percentage of votes cast for Republicans was $\frac{4429}{4429+2950} = 60\%$, and for Democrats it was $\frac{2950}{4429+2950} = 40\%$.

b. In 2010, the percentage of House seats won by Republicans was $\frac{23}{32} = 72\%$, and for Democrats it was $\frac{9}{32} = 28\%$. In 2012, the percentage of House seats won by Republicans was $\frac{24}{36} = 67\%$, and it was $\frac{12}{36} = 33\%$ for Democrats.

c. In 2010, the percentage of House seats won by Republicans was close to the percentage of votes they received. In 2012, the percentage of House seats won by Republicans was higher than the percentage of votes they received.

d. Because of the change noted in part (c), it's plausible that redistricting had an effect on the distribution of representatives.

17. a. In 2010, the percentage of votes cast for Republicans was $\frac{2034}{2034+1882} = 52\%$, and for Democrats it was $\frac{1882}{2034+1882} = 48\%$. In 2012, the percentage of votes cast for Republicans was $\frac{2710}{2710+2794} = 49\%$, and for Democrats it was $\frac{2794}{2710+2794} = 51\%$.

17. (continued)

b. In 2010, the percentage of House seats won by Republicans was $\frac{12}{19} = 63\%$, and for Democrats it was $\frac{7}{19} = 37\%$. In 2012, the percentage of House seats won by Republicans was $\frac{13}{18} = 72\%$, and it was $\frac{5}{18} = 28\%$ for Democrats.

c. In 2010, the percentage of House seats won by Republicans was higher than the percentage of votes they received. In 2012, the percentage of votes received by Republicans decreased only by a small amount, yet the percentage of seats won by Republicans went up by a significant margin.

d. Because of the change noted in part (c), it's plausible that redistricting had an effect on the distribution of representatives.

19. a. The most likely distribution would be 8 Republican and 8 Democrat seats.

b. The maximum number of Republican seats that could be won is 15. In order to get a majority of Republican voters in as many districts as possible, you need to load up one district with Democrats. Each district has 625,000 people – imagine drawing district lines so that all 625,000 people were Democrats in a particular district. That district would elect a Democrat for its representative, but the remaining 15 districts could be drawn to have Republican majorities, and they would elect Republicans. (Note: With the assumption that voting is to take place along party lines, all you'd really need to do is move Democrat from each of 15 districts into the remaining district, and then take 15 Republicans from that district and distribute them evenly across the other 15 districts. This would produce a Republican majority in 15 districts. Of course this assumes that each district begins with exactly 312,500 Republicans, and 312,500 Democrats.) Reverse the above logic to convince yourself that the minimum number of Republican representatives is 1.

21. a. The most likely distribution would be 6 Republican and 6 Democrat seats.

b. The maximum number of Republican seats is 11, and the minimum number is 1. (Refer to Exercise 19b; the logic is the same).

23. a. The most likely distribution would be 15 Republican and 0 Democrat seats. This is due to the fact that with random district lines, one would expect an 80% majority in favor of Republicans in every single district, and as long as voting takes place along party lines, all of these districts will elect Republican representatives.

b. From part (a), we see that the maximum number of Republican seats is 15. In order to find the minimum number, first observe that each district has 500,000 people, and the entire state has 1,500,000 Democrats. A particular district will elect a Democrat only when the district has a Democrat majority. This requires at least 250,001 Democrats (and 249,999 Republicans), with the assumption that voting takes place along party lines (and that everyone votes). However, there are only 1,500,000 Democrats to spread around, and since $1{,}500{,}000 \div 250{,}001 = 5.999976$, the maximum number of districts where one could find a Democrat majority is 5. This implies that the minimum number of Republican seats is 10.

25. There are several solutions. Perhaps the easiest to describe is one where the boundaries divide the state into 8 rectangles, each 4 blocks wide and 2 blocks high.

27. There are several solutions. Perhaps the easiest to describe is one where the boundaries divide the state into 8 rectangles, each 4 blocks wide and 2 blocks high.

29. It is not possible to draw district boundaries that satisfy the conditions given. Three Democrats are needed in a single district for a Democrat majority, and thus a valid solution would require three Democrats in each of three districts, for a total of nine Democrats. But there are only eight Democrats in the state.

31. Answers will vary (there are several solutions).

FURTHER APPLICATIONS

33. Answers will vary for the first three cases (there are several solutions). The last case, where 4 Republicans are elected, is not possible. Five Republicans are needed in a single district for a Republican majority, and thus a valid solution would require five Republicans in each of four districts, for a total of 20 Republicans. But there are only 18 Republicans in the state.

35. Answers will vary.

37. It is not possible for one party to win every House seat for the situation described. To show this for the example provided (20 voters and four districts), note that 3 voters of the same party are required for a majority in a single district. Thus a majority in all four districts would require 12 voters of the same party, which is impossible, because there are only 10 such voters. This argument can be generalized to give a proof of the statement in the exercise.